Transportation Research, Economics and Policy

Feng Xie • David M. Levinson

Evolving Transportation Networks

 Springer

Feng Xie
Metropolitan Washington Council
of Governments
Washington, D.C.
USA
fxie@mwcog.org

David M. Levinson
University of Minnesota
Department of Civil Engineering
500 Pillsbury Drive SE
Minneapolis, Minnesota 55455
USA
dlevinson@umn.edu

Series Editors
David Gillen
Werner Rothengatter

ISSN 1572-4387
ISBN 978-1-4419-9803-3   e-ISBN 978-1-4419-9804-0
DOI 10.1007/978-1-4419-9804-0
Springer New York Dordrecht Heidelberg London

Library of Congress Control Number: 2011925541

Printed on acid-free paper

Springer is part of Springer Science+Business Media (www.springer.com)

For Janet and our newborn son Eric.

Feng Xie

For Samuel, Olivia, and Benjamin.

David Levinson

Preface

I first started thinking systematically about the evolution of transportation networks in the early 1990s when I was a transportation planner at the Montgomery County (Maryland) Planning Department working with the group developing a regional travel demand forecasting model. My job involved developing a new generation of travel demand forecasting models, predicting the number of trips coming into and out of each area, matching trip origins and destinations, determining the mode (car, transit, walking) they would use, and assigning the route on the network they would follow. These models originated in the 1950s in Chicago, and by the early 1990s had been deployed in all large US metropolitan areas. The model I worked on, Travel/2 was distinct in several ways, mainly ensuring an equilibrium between supply and demand (the travel times used to generate travel demand, and the travel times resulting from that demand). However, these models all took as a given the underlying transportation network. If we could forecast traffic growth, and others could forecast land use, why couldn't we forecast the growth of networks? This idea was one of the drivers for me to return to graduate school for a Ph.D.

As I was preparing to return to graduate school in 1994, I remember a conversation I had with my former college roommate Robert Forsythe, who was living and working in nearby Virginia. I proposed that the network itself (not merely the travel times on the network due to the demand, but the physical capacity of the network) responds in some fashion to demand. Like firms, links (segments of the network) with more demand will grow. In the current political arrangement, the decision about which links to expand appears to be made by a central authority. However, it does not have to be that way, and maybe the authority is just responding to local conditions.

I also was thinking about road pricing. It is well understood by transportation economists, but almost no one else, that the only real solution to congestion is pricing: electronically (and wirelessly) collecting tolls from cars, thereby giving travelers the proper signal about the impact of their trip on other travelers. Suppose that instead of a state monopoly on roads, they were privately owned and raised revenue through tolls. What would happen? Would society be better or worse off?

I took those ideas with me to graduate school in Civil Engineering at the University of California at Berkeley, and when I was putting together my dissertation prospectus, I proposed to study *A Model of Road Pricing with Autonomous Links*. Networks would be comprised of independent links, each deciding on a pricing strategy and an investment strategy, given all the others were doing so as well. Each link would charge tolls, and collect revenue proportional to the tolls. Each link would choose a pricing strategy to maximize its profits, conditioned on other links doing the same. I prepared some analytical models, and proposed constructing a simulation framework to look at more complicated cases that could not be analyzed by hand. This proposal did not consider capacity or investment, which it still took as given, though the intent was to relax that assumption.

I proposed this to my committee at my Preliminary Oral Examination in 1996, and in short, though I passed, the prospectus was given a major haircut. The oral examination committee, in its infinite wisdom, thought the topic was too ambitious. One of the members of my dissertation committee said "A monkey can do simulations", which may be true, but a monkey cannot formulate simulations, and I believe simulation is an important tool for understanding how complex networks work.

With the guidance of my advisor Mark Hansen, my dissertation was reshaped into what it became *On Whom the Toll Falls*, which examined historically and analytically the question of whether jurisdictions choose taxes or tolls (which depends on the trade-off between raising revenue from non-residents (favoring tolls) and transaction costs (favoring taxes). The title indicates its consideration of equity and incidence, and the comparison between tax and toll financing regimes. Taxes are cheaper to collect than tolls, but tolls, particularly boundary tolls, can raise money from non-residents much more effectively. This was ultimately published as several papers and a revised version was rolled into the book *Financing Transportation Networks* (Levinson, 2002). Chapter 13 of this extends that dissertation's research by examining which level of government manages particular roads depending on the transportation technology (the speed of travel) and thus the amount of cross-border flows.

I landed at the University of Minnesota in 1999, and picking up the idea that had been set aside, started writing proposals to examine the question of network evolution. In my second year there, I was able to obtain funding to examine the "network growth problem" from the Humphrey Institute's *Sustainable Transportation Applied Research (STAR) Initiative*. The STAR program funded a student, Bhanu Yerra, to develop the models in his Master's Thesis, which became several papers: Yerra and Levinson (2005); Levinson and Yerra (2006). The core essence of that research is replicated in Chapter 8, which lays the groundwork for further simulations.

Another project, funded by the Minnesota Department of Transportation, examining the Twin Cities networks, *If You Build It, Will They Come*, supported the Master's Theses of two other students: Ramachandra Karamalaputi and Pavithra Parthasarathi, and resulted in several more papers: Levinson and Karamalaputi (2003*a*,*b*); Parthasarathi et al. (2003).

The results from these papers were rolled into a proposal to the National Science Foundation: *CAREER: The Evolution of Transportation Networks: Empirical Re-*

search and Agent-Based Models, which was funded (on its third submission) starting in 2003, and supported Feng Xie's dissertation and much of the research presented in this volume.

While suffering a fever at the Regional Science conference in San Juan Puerto Rico in November 2002, I detailed the ideas for the set of papers that became my research agenda for the remainder of the decade. The models outlined then (SOUND, SONIC, SIGNAL) extended the SONG framework described in Bhanu Yerra's thesis, which was being completed at the time.

Other follow-up research included the MnDOT grant *Beyond Business As Usual: Ensuring the Network We Want is the Network We Get*, which funded the research in Chapter 15 (conducted with Norah Montes de Oca).

Finally, I received support from the UK's Economic and Social Research Council to study *The Co-Evolution of Transport Networks and Land Use*, which resulted in Chapters 7, 10, and 12 (the last with Shanjiang Zhu), as well as other related papers (Levinson, 2008*a,b*).

My co-author, Feng Xie, engaged in the network growth research upon arriving in the United States in 2003. This research was his Master's Thesis, Ph.D. Dissertation, and other related research conducted at the University of Minnesota, for which I was the advisor. Except for Chapter 8, this volume is basically a synthesis of his thesis and dissertation chapters and other pieces of network growth studies we have published in peer-reviewed journals and conference proceedings. Without his excellence and commitment to scholarship, this research would not have been nearly as successful.

Feng Xie's original work added an important empirical dimension to the network growth research. He put a significant amount of efforts in extracting historical data and conducting retrospective examination of network growth in empirical cases (Chapters 4-7, Chapter 4 with Michael Corbett). Findings from these empirical analyses not only provide the critical evidence that network growth follows a logical and predictable path, but also shed important light on the subsequent analysis of governance, accessibility, topology, sequence, and land use / transportation interactions.

While SONG (Chapter 8) can be thought of as proof of concept, demonstrating some techniques for modeling the evolution of the hierarchy of road networks, SOUND and SONIC (Chapters 9 and 10) reproduce the observed topological features or sequence of link deployment. Using SONIC/PF and SIGNAL (Chapters 11 and 12) we explored the co-development of transportation networks and land use in simulation, while closely watching the sensitivity of the resulting collective features of networks / land use gradients on model parameters.

Another important contribution of his work, and the main theme of his dissertation, is to explain how ownership affects network growth. Based on the empirical findings (Chapters 4-6) that ownership is a critical factor of network growth and that different ownership structures (and thus different objective functions) may result in different courses of network growth, Xie constructed game-theoretic models in Chapter 13 that associate governance choice with spatial spillovers and network growth. His dissertation work also evaluated the difference in evolution between networks developed under different ownership structures in a controlled environment (Chapter 14).

Engaging in the application of empirically calibrated network growth models to policy and planning studies, Feng Xie also developed a more realistic and sophisticated representation of network growth. On the demand side, he implemented more realistic traffic assignment procedures (stochastic user equilibrium) in travel demand forecasting (Chapter 12); on the supply side, he modeled the investment and pricing policies of suppliers at different levels, considering the benefit, cost, budget, prioritization, variable tolls, inter-jurisdictional collaboration, stated vs. revealed investment rules (Chapters 14 and 15, the latter with Norah Montes de Oca), etc. In addition, it is his original work that used accessibility as an organizing concept to model supply-demand interaction in a holistic model framework of network growth. This economic framework for network growth centered on accessibility distinguishes this line of research from the physics-models that describe how networks might grow without explaining why.

Other students in our Networks, Economics, and Urban Systems (Nexus) research group working on related network growth projects not presented above include Lei Zhang, Wei Chen, and Wenling Chen. The alumni and affiliate researchers of the Nexus group contributed in many ways, and all deserve acknowledgment. Research is in the end a collaborative effort.

David Levinson
Minneapolis
December 2010

Acknowledgments

This book was written based on research work conducted within the project *The Evolution of Transportation Networks: Empirical Research and Agent-Based Models* funded by the *National Science Foundation*.

There are many people that have earned our gratitude for their contribution to this book. We would especially like to thank Michael Corbett, who assembled the Minneapolis skyway data used in Chapter 4, Bhanu Yerra, whose modeling work on the self organization of surface transportation networks has formed the basis of Chapter 8, Shanjiang Zhu, who has helped us with the modeling work reported in Chapter 12, and Norah Montes de Oca, who has helped develop and calibrate the network growth model presented in Chapter 15. We would also like to thank Mr. John Diers for his invaluable suggestions on our streetcar research presented in Chapter 6.

By the time this book is completed, some of its chapters have been written into articles and accepted for publication at journals including *Journal of Economic Geography*, *Transportation Research*, *Public Choice*, *Networks and Spatial Economics*, *Environment and Planning*, and *Computers, Environment, and Urban Systems*. One of the other chapters has been included in *the Proceedings of the 17th International Symposium on Transportation and Traffic Theory (ISTTT)*. We extend thanks to the editors and anonymous reviewers for their inspiring comments and suggestions on our research.

Contents

Part I
ANTECEDENTS

This part introduces the context of this book, overviews the literature, and lays out the framework and main assumptions of the analysis in sequel.

Chapter 1
Introduction

The world would have been radically different without canals, railways, turnpikes, and freeways. Over the last two centuries, the development of these modern transportation systems has been transformative. Evidence abounds. Dating to the late nineteenth and early twentieth centuries when electric street railways were rapidly adopted across the developed world, people relied so extensively on streetcars for commuting and shopping that proximity to a streetcar line determined where they lived and where they worked. As a consequence, the extension of streetcar lines produced finger-shaped residential areas along these lines that featured across many North American cities. Since the 1960s, urban streets, which historically served as both through passage and public places, were swept away and replaced by routes dedicated solely to fast vehicular movement. People did not realize the negative effects of motorized transportation until the road-driven transformation split neighborhoods and jeopardized the sustainability of urban development. Since the infrastructure of most transportation modes takes the form of a network, the temporal change of transportation systems has been characterized by network growth, decline, and structural transformation.

In spite of the profound impact of transportation development, those who have been concerned with its planning and engineering have been accused of giving too little attention to its evolutionary nature. How and why transportation networks change over time? What are the implications of these changes? What are the lessons learnt from the past transportation experience? Answers to these questions are critical not only to create opportunities for better transportation systems in developed countries, but also to provide guidance to the massive infrastructure construction undertaken by developing countries now and in the future. This is why the authors of this book feel strongly about the need to gain an in-depth understanding of the evolutionary growth of transportation networks.

The aim of this book is to understand the process of network growth by identifying and quantifying its determining factors. Transportation development represents a diverse process that involves a magnitude of dimensions, which may be topological (e.g., addition and abandonment of linkages between places and facilities), hierarchical (e.g. the importance given to some roads or facilities over others), morphological (e.g., structural changes to infrastructure in shape and orientation), temporal (e.g., historic legacy, lock-in, forecasting, planning), technological (e.g., emerging transportation modes, advances in intelligent transportation technology), economic (e.g., tax or toll, nationalization or privatization), managerial (e.g., regulation, signal control, car-pooling), social (e.g., congestion, pollution, sprawl, "highway mania"), or political (eg., policy-making, equity, funding allocation). Under the umbrella of network growth, this book focuses on a narrow set of economic and regulatory factors which, while being revealed to be vital to network growth, have not been fully understood or modeled in the past. While analysis is primarily based on the past transportation experience of the United States, this book addresses contemporary needs of transportation planning, including important considerations such as environmental consequences and the larger macro-economy, and to promote positive future transportation development around the world.

Over the last two decades, some large developing countries such as Brazil, China, and India have made enormous investments in transportation infrastructure in the process of rapid economic development. Over one decade from 1990 to 2000, the total length of motorways in China has dramatically increased from 500 kilometers to 16,314 kilometers, while the total length of road infrastructure in India has grown by over 50% over the period of 1980 to 2002 (International Road Assessment Programme, 2008). How networks of transportation infrastructure can be designed and deployed to efficiently serve the booming economic needs is a subject of intrinsic interest for the ongoing policy-making efforts in these countries.

In contrast with developing countries, the United States anticipates little growth over the next several decades in the networks that currently serve freight and passenger travel needs. In fact, since the completion of the Interstate Highway System in the 1980s, focus has shifted within transportation agencies from large-scale capital-intensive investments to the improved operations, maintenance, and management of the mature system. In response to the growing concerns for congestion, environment, and limited funding resources associated with the aging infrastructure, the last decade has witnessed significant attention to the concept of sustainable transportation, which aims to move people, goods, and information efficiently while reducing the negative impact on environment, economy, and society. In spite of the stagnant supply, transportation demand still expands with population and trade, especially within fast-growing metropolitan regions. Transportation networks that serve growing travel demand have become increasingly interconnected at a regional level. Recent years have seen a widespread call for better coordination between jurisdictions, departments, and private agencies involved in the management, operation, and financing of these networks. Central to the efforts to respond to this call is the establishment of Metropolitan Planning Organizations (MPO) and various Regional Operating Organizations (ROOs), which mark a significant shift of institutional transportation decision making from smaller municipalities and larger states to a metropolitan level.

Indeed, the effective investment and operation on transportation infrastructure networks has shaped the United States' national priorities for economic development, homeland security, and energy supply. How best to plan, manage, and regulate sustainable transportation development under interdependent economic and regulatory initiatives challenges transportation policy-makers and professionals. Conventional planning approaches with a primary focus on vehicular mobility within a top-down decision-making framework, however, do not adequately account for the interconnected and interdependent nature of current transportation systems, and may produce socially and economically undesirable outcomes in the long run. To bridge this gap, the concept of evolutionary transportation planning has gained recognition in recent years, attempting to address how individual decisions and actions could eventually accumulate into development processes which are both path dependent and unpredictable. This book uniquely contributes to this development by examining the evolutionary growth of transportation networks in explicit consideration of interdependent economic / regulatory interests in decentralized investment processes.

The book is organized into six parts. Part I introduces the context, overviews the literature, and lays out the framework of our analysis. Part II presents empirical studies that explore the deployment of transportation networks using historical data. Part III constructs analytical explanations of network growth by demonstrating the spontaneous organization of network geometries including hierarchy, topology, and sequence in a series of *ex ante* models. Part IV takes a look at the co-evolution of transportation networks, places, and the distribution of land use activities. Part V applies empirically calibrated network growth models to transportation planning and policy studies from a normative perspective. The topics in examination include governance choice and forecasting. Each of the chapters in Part II - Part V can be viewed as a stand-alone study. Part IV highlights the findings from these studies, and discusses their implications for evolutionary transportation planning.

Chapter 2
Background

The city of Königsberg, Germany was set on both sides of the Pregel River. As shown in a sketch map of the city (Figure 2.1(a)), seven bridges across the river connected two large islands and the mainland to one another. The people of Königsberg wondered whether or not one could walk around the city in a way that would involve crossing each bridge exactly once. Euler approached this problem by collapsing areas of land separated by the river into points (Figure 2.1(b)), and representing bridges as curves (Figure 2.1(c)). In modern graph theory, points are often referred to as vertices and curves as edges. Euler observed that (except at the endpoints of the walk) whenever one enters a landmass by a bridge, one leaves the landmass by a bridge. In other words, during any walk in the graph, the number of times one enters a non-terminal vertex equals the number of times one leaves it. Now if every edge is traversed exactly once (except for the ones chosen for the start and finish), the number of edges touching a vertex is even (half of them will be traversed "toward" the vertex, the other half "away" from it). However, all four land masses in the original problem are touched by an odd number of bridges (one is touched by 5 bridges and the other three by 3). Since at most two land masses can serve as the endpoints of a putative walk, the existence of a walk traversing each bridge once leads to a contradiction.

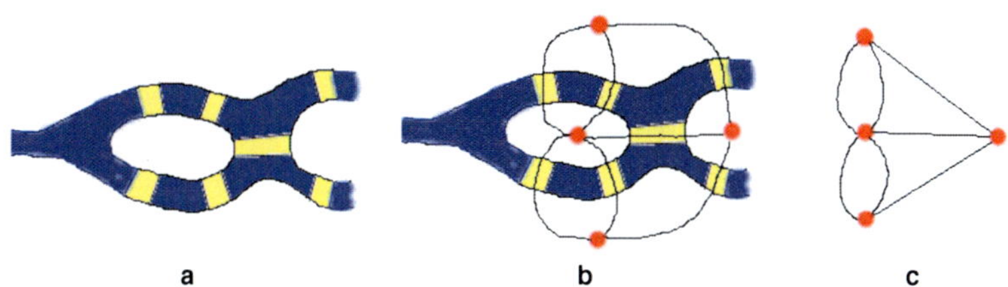

Fig. 2.1 The seven bridges of Königsberg. Source: Math Forum (2010)

The classical problem of the seven bridges of Königsberg introduced perhaps the first transportation network in a scientific sense. Since then there has been a long-established interest for scholars and professionals in gaining a greater understanding of transportation networks. Of particular interest for this book is the temporal changes of transportation network systems. The literature that discusses this subject has been prolific. Three comprehensive historical reviews, for example, came from Fullerton (1975), who introduced the development of British transportation networks, Taaffe et al. (1996), who outlined the evolution of the U.S. transportation systems, and Garrison and Levinson (2006), who examined transportation experience in the past centuries from the perspectives of transportation policy, planning, and deployment.

Despite the fact that the growth of transportation networks is complicated and multidimensional, and its duration is usually measured in decades, it may still be tractable and predicable with a further understanding of its underlying mechanisms. Under this belief, sustained efforts have been put in the modeling and analysis of transportation networks in a range of fields from geography, regional science, economics, network science, and urban planning, to transportation engineering.

Previous studies have mainly followed five streams.

- In the 1960s and 1970s, geographers viewed network growth as topological transformation, aiming to either extract the process of structural changes or replicate the emergent topologies of transportation networks.
- Since the 1970s, the prevalence of travel demand forecasting models has provided transportation planners and economists with an effective tool for predicting traffic flows on a network and modeling the optimal changes to the network, with the belief that network growth is the result of rational decisions by jurisdictions, property owners, and developers in response to market conditions and policy initiatives.
- Recent large-scale statistical analyses, made possible by the availability of sufficient historical geographic data and increasing data processing ability, related the temporal change of transportation supply (the presence or absence of infrastructure, change in service frequency or capacity, etc.) to the demographic and socioeconomic characteristic of tributary areas, as well as traffic conditions and other attributes of infrastructure.
- The economics of network growth had examined the formation and growth of a wide array of networks, and explored the growth of transportation networks in particular from various perspectives such as traditional transportation economics, urban economics, fiscal federalism, public economics, network effect, path dependence, and coalition formation.
- Since the "new network science" came onto the scene in the 1990s, interest has emerged in introducing the concepts of preferential attachment and self-organization, and the technique of agent-based simulation to model the evolution of a transportation network as a spontaneous process, by which independent initiatives and behaviors of individual travelers, providers, and regulators play out collectively into transportation development.

This chapter surveys this substantial body of studies with a particular focus on modeling and quantitative analysis. The next five sections introduce the five main streams of studies. The last section summarizes their subjects, methods, and connection rules.

2.1 Transportation geography

Transportation networks are commonly simplified as graphs with elementary components retained: nodes indicate centroids of human settlements (places), facilities, and intersections of routes; links represent segments of infrastructure or service routes; flows represent the actual patterns of movement on networks. It was not until 1962 that Garrison and Marble (1962) introduced graph theory to the study of transportation networks in the fields of geography, regional science, and urban studies (Lowe and Moryadas, 1975). During the heyday of the economic geography / regional science movement in the 1960s and 1970s, a few studies were conducted by geographers to model the growth of transportation networks in terms of their structural transformation and topological changes. The most comprehensive outlines of these are found in Haggett and Chorley (1969) and in Lowe and Moryadas (1975).

Attempts have been made to model the continuous growth of transportation networks in a series of discrete stages. Taaffe et al. (1963) proposed a four-stage model to describe the sequential process of road network development when colonial exploitation proceeds from the coastal baseline to the inland area in an underdeveloped country. The model is illustrated in Figure 2.2, where dots represent places while lines represent roads that connect these places. Size of dots and boldness of lines represent relative scale of places and roads, respectively. As shown in Figure 2.2, scattered ports with equally small size are located along the coast of a colonial region (A); later on, penetration lines are built from interior to reach selected ports (B); connected ports and inland feeders then develop because of the growth of inland trading (C); as more links are built to interconnect developed nodes, links are also differentiated and important links emerge (D). Pred (1966) applied the Taaffe model to Atlantic seaboard of the United States, while Rimmer (1967) applied the model to the South Island of New Zealand. Lachene (1965) developed a staged model of network development on a hypothetical isotropic transportation network. The model starts with a network of dirt trails and a more or less uniform distribution of economic activity. As towns form at some intersections, a road network is built to link these settlements. While some trails become paved roads, some less used links are abandoned in the countryside when economic activities concentrate in the urban centers. Finally a superior network, perhaps a railroad or a freeway, emerges connecting the urban centers.

Another strand of geographical studies, rather than describing network growth in stages, constructed models that would replicate observed network patterns. Garrison and Marble (1962) described their attempts to simulate the changing topology of the Northern Ireland railroad system between 1830 and 1930 using Monte Carlo sim-

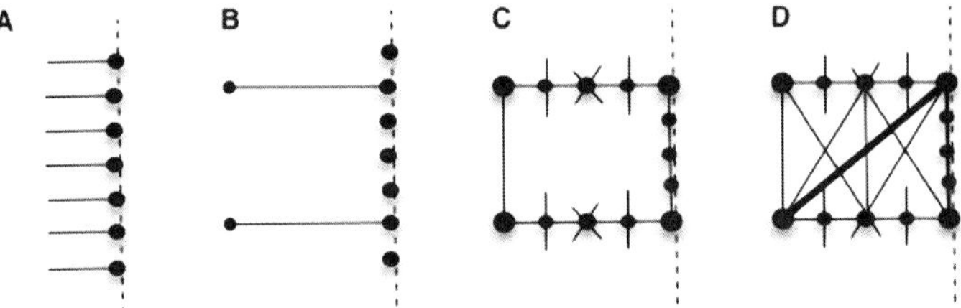

Fig. 2.2 An illustration of the four-stage model of network growth

ulation methods, while Morrill (1965) reported parallel studies on the rail network of central Sweden. Kansky (1963) developed a quantitative predictive model of network structure and applied it to the Sicilian railroad. He selected 16 settlements in Sicily which by 1908 would have been on the railroad network in a random process among 30 major settlements in this region. The first link is added to connect the two largest centers and then links are gradually added such that the next largest center joined the largest and closet center in the network. Kolars and Malin (1970) modeled the development of the Turkish railroad network employing an approach different from node connection, by which transportation links emerge on major ridge lines as defined by a population-accessibility surface. Black (1971) conceived of the railroad network in Maine as a tree branching out from Portland, growing outward to connect outlying peripheral nodes. The possibility of constructing a link is calculated as a function of potential revenue, construction cost, with a constraint of the angle of the link. The presence or absence of a link between a pair of vertices at a particular time is determined by whether the score exceeds a threshold.

While these studies provide insight into the structural change of transportation networks, they had to deal with simple networks using heuristic and intuitive rules for network growth and transformation, due to the lack of understanding on the inherent mechanisms with regard to why and how transportation networks evolve. Based on a review of these contemporary studies, Haggett and Chorley (1969) found that they were "somewhat fragmentary", and suggested that "a general theory of network growth lies in future research". The study of the growth of transportation networks remained largely dormant for the following thirty years.

2.2 Optimization and network design

In the late 1950s, Dr. J. Douglas Carroll, Jr. initiated an effort in the Chicago area to seek an optimal solution to investment in roads in terms of minimizing the total cost of road investment, users travel time, operating costs and accident costs. He and his team devised a sequence of forecasting methods to forecast the amount of personal travel that would occur on a typical weekday in 1980 by road and transit, and derived an expressway spacing formula to work out the extent and layout of the system of expressways and arterial roads to serve the existing and perspective pattern of urban activities. Boyce (2007) provides an account of this pioneering effort. The Chicago Area Transportation Study (CATS) soon became known worldwide as the premier example of travel demand forecasting for urban areas (Sheffi, 1985; Ortuzar and Willumsen, 2001). The prevalence of travel demand models enabled the prediction of traffic flows on transportation networks in a realistic way, and led to the marked revival of the interest in modeling the evolution of transportation networks.

Traffic flow plays an essential role in driving network growth. Newell (1980) and Vaughan (1987) examined traffic flows in shaping various network geometries. From a more practical perspective, Ewing (2000) proposed a sketch planning methodology to determine the optimal spacing of through streets that accounts for changes in mode share, trip length, time of travel, and intersection capacity as residential density increases. Vitins and Axhausen (2010) explored the generation of optimal network designs from a featureless plane using pre-existing network patterns and grammars.

In the last two decades, solution algorithms to user equilibrium enable the derivation of traffic flows across transportation networks, and have been widely incorporated to solve the network design problems (NDP) (LeBlanc, 1975; Yang and Bell, 1998). Typically the NDP is formulated as a bi-level framework in which the lower-level represents the demand-performance equilibrium for given investment while the upper level represents the investment decision-making of the transportation planner to maximize social welfare based on the unique equilibrium flow pattern obtained from the lower-level problem. A continuous NDP problem deals with the optimal capacity expansion of existing links while a discrete NDP derives the design of an optimum amount of transportation supply by changing the actual topology of the network, that is, by adding or removing links. Constrained by computational ability, the choice set of discrete changes has been limited to a small size.

If the NDP were how decisions are made, network changes would be due to planners' rational behaviors to maximize the efficiency of a given network, which has to be measured according to some quantifiable objective based on predicted traffic with budgetary and other constraints. When various factors come into play in reality, however, objectives may be ambiguous and decisions may be made with uncertainty. Curry (1964) claimed that while every locational decision may be optimal from a particular point of view, the resulting actions as a whole may appear to be random. Because of spatial lock-in, transportation networks may have a locational stability which is greater than individual components making up these networks. From the perspective of transportation economics, Zhang and Levinson (2007) argued that the NDP perspective simplified a network growth problem in three aspects: top-down investment decisions are considered independent of decentralized pricing and regulatory structures; only optimal investment rules are considered; and inter-dependencies of sequential decisions are ignored. Bertolini (2007), observing that conventional planning approaches do not adequately account for the irreducible uncertainty of future developments, proposed an evolutionary approach of urban transportation planning to address how individual decisions and actions could eventually accumulate into development processes which are both path dependent and unpredictable.

2.3 Empirical models of network growth

In the past, although statistical analysis had been widely used in regional science and transportation studies, it found limited application in analyzing the growth of transportation networks, largely due to the scarcity of historical data. Not until the last decade, with the availability of sufficient data and increasing data processing ability, especially powered by Geographic Information Systems (GIS), cross-sectional time-series analysis has seen widespread application to investigate the temporal change of transportation supply based on historical observations on a regional level.

The relationship between transportation supply (network) and demand (land use) has been widely examined as a two-way process by which one is the driver of the other. Early work by Gaudry (1975) and Alperovich et al. (1977) employed simultaneous equations to examine the mutual causality of transit demand and supply. Extending their research, Peng et al. (1997) developed a simultaneous route-level transit ridership model and estimated the model using the data from the Tri-County Metropolitan Transportation District of Oregon (Tri-Met) service area. The results indicate that simultaneity exists between transit demand and supply, especially the service supplied is influenced by the past ridership and current demand. Taylor and Miller (2003), on the other hand, accounted for the simultaneity between transit demand and supply using a two-stage least squared (2SLS) regression method. Cervero and Hansen (2002) employed a simultaneous equations system to estimate both vehicle miles traveled and lane miles of supply, suggesting that the relationship is two way, and that similar forces are at work affecting changes in both travel demand and infrastructure supply. Levinson (2008a) examined the mutual causality between the changes that occurred in the rail network and density of population in London. With panel data representing the 33 boroughs of London from 1841 to 2001, models were estimated using the panel corrected standard errors (PCSE) procedure and results disclosed the spatial co-development of rail networks and population in London. Levinson and Chen (2005) employed a Markov Chain Model to analyze the spatial co-evolution of transportation and land use for the Minneapolis-St. Paul Metropolitan Area from 1958 to 1990. A transition matrix records the interaction between transportation and land use and is used to predict the future development of transportation and land use.

Treating land use as exogenous, Mohammed et al. (2006a) explicitly modeled the temporal change in transit supply. In their attempt to model the changes that occurred over a 15-year period in the bus network of the City of Missisauga, Toronto, they employed multiple regression and simultaneous equation models to relate transit supply (measured by bus frequency) to a group of demographic, socioeconomic, and route-specific variables. In a subsequent study, they further introduced artificial intelligence to account for the behavior of transit agencies and simulate the growth of transit routes (Mohammed et al., 2006*b*). At a microscopic level, Levinson and Karamalaputi (2003*a,b*) examined in two parallel studies the expansion and new construction of a network, respectively, based on the present conditions of the network, traffic demand, demographic characteristics, project costs, and a budget constraint. Binary logit (and mixed logit) models were used to associate the expansion and new construction of each link with historic data including physical attributes of the network, their expansion and construction history and AADT values on each of the links. Levinson and Chen (2007) developed an area-based model of highway growth; binary logit models were adopted to estimate the new route growth probability of divided highways and secondary highways using high-quality GIS data of land-use, population distribution, and highway network for the Twin Cities Metropolitan Area from 1958 to 1990.

Another stream of empirical studies was focused on the correlation between the development of transportation networks and changes in collective network features. Blumenfeld-Lieberthal (2009), for instance, measured the concentration of modern air and rail transportation networks in Germany, Italy, Poland, the United Kingdom, and the United States and associated the connectivity of the network (measured by a clustering coefficient) with economic growth and GDP. Erath et al. (2009) investigated the development of the Swiss road and railway network during the years 1950-2000 using an array of topology, centrality, and local efficiency measures, and showed that the freeway network has become less tree-like (and thus more connected) over time, and similarly has become more efficient over this period. El-Geneidy et al. (2011), examining the historical growth of Montréal's indoor pedestrian network, calculated changes in the level of access to retail space, and indicated access to retail and public transit had a major impact on the growth of the Indoor City. Atack and Margo (2011) studied the impact of access to rail transportation on agricultural improvement using the American Midwest in 1850-1860 as a Test Case. Using a GIS-based transportation database linked to county-level census data, they estimated that at least a quarter of the increase in cultivable land can be linked directly to the coming of the railroad to the Midwest.

2.4 Economics of network growth

Economists have examined the formation and growth of a wide array of networks (social, industrial, tele-communication, physical infrastructure, etc.) from various perspectives of supply, demand, benefit, cost, financing, regulation, information, externalities, and so on. While transportation economics has widely introduced traditional supply-demand theories to address contemporary transportation issues, there are other aspects of economics that have been revealed to be essential to network growth, but have not yet seen application to the modeling and analysis of transportation networks. Examples include urban economics, fiscal federalism, network effect, path dependence, and coalition formation. This section surveys these branches in turn.

2.4.1 Transportation economics

Transportation economists have introduced traditional supply-demand theories to investigate a wide range of transportation issues such as congestion, pollution, road pricing, and project evaluation (Gómez-Ibáñez et al., 1999). For the purpose of this book, of particular interest is the investment in transportation infrastructure networks under alternative pricing and regulatory regimes. Quantitatively, this issue has been approached from two directions. Theoretical exploration focused on the endogenous choice of prices, investment, and ownership on transportation networks. Because of the computational complexity associated with large-size networks, this analysis has been largely focused on small networks with one Origin-Destination pair and two or more alternative routes (Button, 1998; de Palma and Lindsey, 2000; Verhoef et al., 1996; Verhoef and Rouwendal, 2004; Borger et al., 2005; Zhang and Levinson, 2007) or with stylized network geometries (Levinson, 2002). Empirical studies, on the other hand, were able to examine the financing of transportation networks of larger scales using econometric methods, with a primary focus on the relationship between the performance of transportation systems, economic / demographic characteristics of tributary regions, and the allocation of funds or decision-making power across regulatory hierarchies (Humplick and Moini-Araghi, 1996*b*,*a*; Levinson and Yerra, 2002).

2.4.2 Urban economics

A stream of urban economic studies has examined the evolution of urban space including transportation as one of the determining factors. The pioneering work by von Thünen (1910) presented a monocentric city surrounded by agricultural land and predicted the rent and land use distribution for competing socio-economic groups. Christaller (1933) introduced Central Place Theory and demonstrated that a hierarchy of central places will emerge on a homogeneous plain to serve the surrounding market while minimizing transportation costs. Krugman (1996) explored the phenomenon of self-organization in urban space. He developed an edge city model to demonstrate how interdependent location decisions of businesses within a metropolitan area could lead to a polycentric pattern under the tension between centripetal and centrifugal forces. Based on these theoretical investigations, a host of empirical land use-transport models have been developed to forecast land use development while considering transportation as an important factor. One of the first that gained substantive interest was the Lowry model (Lowry, 1963). Since the 1980s, many integrated land use models have seen applications in urban planning for real cities and some have been developed into commercial packages. Examples include START (Bates et al., 1991), LILT (Mackett, 1983, 1990, 1991), and URBANSIM (Alberti and Waddell, 2000). Comprehensive reviews of these integrated land use-transport models have been provided by Timmermans (2003) and Iacono et al. (2008). As evidenced by both theoretical and empirical studies of urban development (Hansen, 1959; Guttenberg, 1960; Huff, 1963; Murayama, 1994; Ahlfeldt and Wendland, 2010), the dynamics of urban space has been played out as the outcome of the location decisions made by residents, developers, and business owners, in which accessibility to land use activities such as employment and residence play essential roles.

2.4.3 Fiscal federalism

It has long been observed that the development of all transportation modes has been affected by "a constantly shifting mix of laissez-faire economics stressing private enterprise, on the one hand, and government initiatives at local, state, and national levels, on the other hand" (Taaffe et al., 1996). The organization of ownership has been essential in shaping transportation networks over time. In particular, the governmental provision of general public goods (including roads and other transportation infrastructure) in a federation has been a classic problem examined by a broad literature of financial federalism (Oates, 1972; Besley and Coate, 2003). A branch of this literature focuses on the discrete choice between centralized versus decentralized provision of public goods (Epple and Nechyba, 2004). In general, centralized provision of transportation infrastructure involves a single unitary government that is responsible for the financing, investment, maintenance, and operation of transportation networks (e.g., roads), while a decentralized pattern involves autonomous local jurisdictions that spend on networks independently. The fiscal federalism literature will be discussed in greater details in Chapter 13.

2.4.4 Network effect

The paper "The Economics of Networks" by Economides (1996) opened the way to examine a salient collective property of network industries, called alternatively a network externality or network effect. The network effect has been defined as a change in the benefit, or surplus, that an agent derives from a good when the number of other agents consuming the same kind of good changes. Since this type of side effect is known as an externality in economics, externalities arising from network effects are known as network externalities. Positive network effects are obvious as one's value increases from others' using the same product, while negative network effects also exist, especially where there are resource limits, such as on the overloaded freeway or crowded bandwidth. Shapiro and Varian (1998) illustrated network externalities using a simple demand and supply model, which predicts a typical process of network growth under network effect: the number of users connected to the network is initially small, and increases only gradually as costs fall; but when a critical mass is reached, the network growth takes off dramatically.

Network externalities have played a fundamental role in driving network dynamics in many network industries such as telecommunications, financial exchanges, software, and the Internet. Transportation, as a network industry, is no exception. When a new place or facility with residents and businesses is connected to a transportation network, the residents and businesses at other already connected places benefit from the new connection because they now enjoy accessibility to more activities. Evidence of network effects in transportation networks has been provided by studying the history of transportation networks. Nakicenovic (1998), by plotting a large number of curves for transportation systems, showed that S-curves fit the temporal realization of transportation networks very well. As suggested by the S-curves, there is a long period of birthing as the network is researched and developed, a growth phase as the network is deployed, and a slower mature phase as the network has occupied available market niches. For instance, the US highway network expanded very slowly from 1860 to 1920 and from then on started an exponential climb, before slowing again by the end of the twentieth century. These observations coincide with the prediction of Shapiro and Varian (1998)'s network growth model subject to network effect.

In a historical case study, Bogart (2009) examined how network externalities affected the growth roads, canals, and ports during the early English Industrial Revolution. The author found that there are positive inter-modal network externalities, and that the presence of roads increased the development of canals, as the necessary local feeder road network may have reduced the risks associated with canal development. Negative network externalities, on the other hand, may arise from decentralization and excessive competition between suppliers. Casson (2009), after studying the evolution of the British railway network during 1825-1914, claimed that the network was over-capitalized with excessive duplication of lines due to over competition between towns which the national government was too weak to control.

2.4.5 Path dependence

Today's transportation systems result from what happened in the past, thus the current conditions of a system have a sensitive dependence on its initial conditions, which is referred to as "path dependence" (Arthur, 1994). Liebowitz and Margolis (1995) indicated where information is imperfect, a certain form of path dependence may lead to lock-ins and market failure that are regrettable but costly to change, even in a world characterized by independent decisions and individually maximizing behavior. Despite an increasing realization that transportation development is a sequential process which clearly does not follow a socially optimal design, due to imperfect information when local or individual "optimal" decisions are made (Bertolini, 2007; Zhang and Levinson, 2007), how and to what extent imperfect information and path dependence affect network growth remain unclear to scholars.

2.4.6 Coalition formation

How groups form and are organized to conduct political, economic, and social activities are subject to intense game-theoretic research (Demange and Wooders, 2005). The seminal work by Jackson and Wolinsky (1996) aroused a new stream of contributions using networks (graphs) to model the formation of links among individuals. Marini (2007) provides an overview of recent developments in the theory of coalition and network formation for economic applications. These advances in economic theory also shed new light on the research of network growth. Transportation development cannot be divorced from the interplay of independent jurisdictions and (or) private companies that have provided transportation infrastructure at local, regional, and national levels. Taking the Interurban network of Indiana for example, the network had been constructed and operated by more than 20 private firms (Hilton and Due, 1960). As another example, Virginia State Route 267 consist of three sections (two toll roads of the Dulles Toll Road and Dulles Greenway and a free road for Dulles Airport access) that are operated by Virginia Department of Transportation, Toll Road Investors Partnership II (TRIP II), and the Metropolitan Washington Airports Authority, respectively. How jurisdictional or industrial providers and operators develop a transportation network in a joint process also deserves academic examination.

2.5 Network science

Traditionally, network scientists modeled the dynamics of a transportation network in the attempt to extract or generate the optimal structure of networks. Schweitzer et al. (1998), for example, investigated the evolution of road networks during the optimization process by which a minimized travel detour is traded-off against a minimized cost of constructing and maintaining roads. Gastner and Newman (2006) presented an optimization model to minimize the cost of building and maintaining a transportation network. Optimized network structures were able to replicate the qualitative features of the networks with or without spatial constraints, with one parameter in the cost function varied. Barthélemy and Flammini (2006) proposed a model of traffic networks via an optimization principle. The topology of the optimal network turns out to be a spanning tree and, by changing model parameters, different classes of trees are recovered. Adamatzky and Jones (2009), recognizing the similarity between road planning and plasmodium's behavior to span spatially distributed sources of nutrients with a protoplasmic network, studied the optimal layout of transport links between the ten most populated urban areas in United Kingdom from the "plasmodium's point of view". Simulation results show that during its colonization of the experimental space the plasmodium forms a protoplasmic network isomorphic to a network of major motorways except the motorway linking England with Scotland. In another effort to study biologically inspired adaptive network design, Tero et al. (2010) showed that the slime mold Physarum polycephalum forms networks with comparable efficiency, fault tolerance, and cost to the Tokyo rail system.

Since the 1990s, scientific interest in the structure of complex networks has been aroused by the observation of a power-law distribution in a variety of so-called "scale free" networks, such as the World Wide Web, metabolic networks, citation networks, and the network of human sexual contacts (Albert et al., 1999; de Solla Price, 1965; Jeong et al., 2000; Liljeros et al., 2001). Newman (2003) presented a comprehensive review of the literature. The book *Linked: The New Science of Networks* by Barabási (2002) popularized the new network science emerging from these findings.

As the physics community became interested in surface transportation networks, however, it was recognized that some networks exhibit topological attributes that differ from the typical "scale free" networks. Notable examples are networks with strong geographical constraints, including power grids and surface transportation networks. Csányi and Szendröi (2004) demonstrated a clear dichotomy between large real-world networks which are small worlds with exponential neighborhood growth, and fractal networks with a power-law distribution. Typical examples of the latter are networks with strong geographical constraints, including power grids and surface transportation networks; Gastner and Newman (2006), revealing that the structure of geographical networks are distinct from non-geographical ones, provided a connection between the two classes of networks in that they both can result from the same optimization model with one parameter varied. Specifically, De Montis et al. (2006) studied the interurban commuting network of the Sardinia region in Italy, and disclosed that the statistical properties of traffic structure exhibit complex features and non-trivial relations with the underlying topology; Jiang and Claramunt (2004); Jiang (2005) and Jiang (2007), after analyzing the street-street intersection topology (in which all named streets are represented as nodes, while street intersections as links) of urban street networks across North America and Europe, found that urban street networks exhibit a scale-free property characterized by a connectivity distribution with a power law regime followed by a cutoff. Derrible and Kennedy (2010), by looking at 33 metro systems in the world, found that most metros are indeed scale-free and small-world networks, but they show atypical behaviors with increasing size. Barthélemy (2010), reviewing the most recent empirical observations and cutting-edge models of spatial networks, investigated how spatial constraints affect the structure and properties of these networks.

In exploring how scale-free networks emerge and evolve, Barabási and Albert (1999) found that as new nodes enter a scale-free network, they are more likely to link to highly connected nodes than lesser connected nodes, and this feedback loop gives preference to the large nodes. They called this process "preferential attachment", which has been intensively studied to explain the dynamics of complex networks (Jeong et al., 2000; Barabási, 2002; Dorogovtsev and Mendes, 2002). Although this "rich get richer" growth mechanism of preferential attachment does not seem to perfectly apply to transportation networks due to geographical constraints, it provides some insight to transportation studies: First, preferential attachment may explain the emergence of hub-and-spoke systems in less constrained transportation networks such as airline and shipping networks. Second, when independent nodes link to a network, they tend to connect to established and more important nodes, although the importance of a node is not necessarily associated with the number of connections as it is in a scale-free network, and the direct connection may be realigned to reduce cost and avoid competition between redundant routes. Third, large-scale order and organization may emerge in transportation networks based on independent decisions, which has been extensively examined in another emerging scientific field, self-organization.

Self-organization exists in many complex systems that seem to spontaneously evolve into large-scale order, even based on simple behaviors of independent agents in the systems (Schelling, 1978). Since the late 1990s this concept has been introduced to interpret the evolution of various complex networks ranging from the Internet and social networks, to biological networks employing simulation methods (Newman, 2003). A branch of these studies modeled the emergence of morphologies and patterns in cities (Batty and Xie, 1994; Krugman, 1996; Samaniego and Moses, 2008; Courtat et al., 2010). In recent years, agent-based simulation has seen applications to interpret the dynamics of transportation networks. Lam and Pochy (1993) and Lam (1995) proposed an active-walker model (AWM) to describe the dynamics of a landscape, in which walkers as agents moving on a landscape change the landscape according to some rule and update the landscape at every time step. Helbing et al. (1997) adopted the active walker model to simulate the emergence of trails in urban green spaces shaped by pedestrian motion. In this process, pedestrians directly walked to their respective destinations on a homogeneous ground at the beginning. Then frequently used trails got reinforced since they are chosen by pedestrians more while rarely used trails withered and were finally destroyed. Consequently, the trails bundled and emerged into different patterns. Helbing et al. (1997) found out that their model was "able to reproduce many of the observed large-scale spatial features of trail systems." Yamins et al. (2003) present a simulation of road growing dynamics on a land use lattice that generates global features as beltways and star patterns observed in urban transportation infrastructure. However, their simulation did not consider the dynamics of traffic flows. Zhang and Levinson (2004) examined the growth of a real-world congesting network - the Twin Cities (Minneapolis / Saint Paul) road network with autonomous links. Based on the network topology in 1978, simulation experiments were carried out to "predict" road expansions in twenty years, and the predicted 1998 network is compared to the real one. Yerra and Levinson (2005) and Levinson and Yerra (2006) demonstrated that a road network can differentiate into an organized hierarchical structure from either a random or a uniform state, suggesting that the hierarchy of roads, rather than necessarily being designed by planners or engineers, is an emergent property of network dynamics. Based on a principle of local optimality, Barthélemy and Flammini (2009) developed a simple model of formation and evolution of city roads which reproduced the most important empirical features of street networks in cities.

2.6 Summary and discussion

The temporal development of transportation systems such as inland waterways, turnpikes, rails, airlines, and roads over the last two centuries is complicated and multidimensional. The particular focus of this review is on the modeling and analysis of the growth of transportation networks. Efforts over the last half-century have aimed to model network growth in a broad range of fields including physics, geography, economics, natural science, urban planning, and transportation engineering, and not surprisingly, generated a wide literature that varies in subject, method, and growth mechanism. Table 2.1 summarizes a selection of these studies in chronological order for an overview.

As can be seen in this table, geographers in the early days had to limit their modeling efforts to heuristic and intuitive connection rules that allow them to replicate the observations of structural changes in networks, due to the lack of understanding of underlying growth mechanisms. It was not until the introduction of travel demand modeling and formal models of user equilibrium researchers were able to predict traffic flow across a network in a systematic way, thereby solving the "optimal" changes in transportation supply that minimize user cost on the network under budgetary constraints. Since then the concept of a bi-level optimal network design has dominated the decision-making mindset in urban transportation planning.

In contrast to the static, one-dimensional environment in which optimal network designs were solved, economists reveal a more complex world. In reality, transportation development has been the accumulative outcome of individual decisions that are made from independent economic and political initiatives. Organization of strategic providers from public vs. private and local vs. regional interests may significantly affect the course of network growth. Transportation development also demonstrates characteristics such as network effects and path dependence due to externalities and incomplete information that arise from an evolutionary process. These dimensions, however, have not been formally treated in previous network growth models. Time-series statistical analyses that relate the changes in transportation supply to various demographic, economic, and technical factors based on historical observation provide some insights, although they are costly and largely post hoc and case-specific. Thus it remains a challenge for researchers and practitioners to develop a systematic evolutionary approach of transportation planning by which transportation development could be modeled in a more realistic way and transportation supply could be provided more effectively.

Recent scientific advances in modeling complex systems and complex networks provide new opportunities. Tremendous interest has been aroused to interpret the growth of transportation networks adopting the concepts of preferential attachment and self-organization from natural science. Agent-based simulation has provided an effective tool by which independent initiatives, behavioral rules, and travel demand forecasting could be included in an integrated process, and has seen widespread applications in network growth models. One caveat is that due to the complexity of the issue examined, most agent-based simulations were only able to include simple and myopic objectives or behavioral rules of agents. Obviously realism of models could be improved at the cost of adding more complexity, but at which point the trade-off between realism and complexity should be made remains an open question. Additionally, most of these models have been exploratory, in the sense that theories are presented without validation and empirical models are validated only on some basic aggregate features (e.g., visual geometric similarity). Is this evidence sufficient to support a model of such a complex process? How far can scholars and practitioners go with these network growth models to predict the future, or at least provide insightful implications for transportation planning? These questions deserve further investigation.

While there have been some investigations of the evolution of transportation infrastructure and that of urban land use separately, few have examined the integrated development of transportation and urban space from an evolutionary perspective, leaving the co-evolution of transportation and land use still poorly understood.

Table 2.1: Summary of selected studies on network growth

Reference	Market	Given	Rules	Method
Garrison and Marble (1962)	Railroads in Ireland	Nodes	Connect to nearest neighbor	Heuristic
Taaffe et al. (1963); Pred (1966); Rimmer (1967)	Underdeveloped road networks close to costal line	Emerging ports, inland establishments	Staged node connection	Heuristic
Lachene (1965)	Idealized	Grid of dirt trails	Staged node connection	Theoretical
?	Sicilian railroads	Major settlements	Quantitative predictive	Empirical
Black (1971)	Maine rails	Places	Potential scoring of links	Heuristic
LeBlanc (1975)	Sioux Falls	Network, budget	DNDP	Theoretical
Lam (1995); Lam and Pochy (1993)	Idealized	Homogeneous space	Active Walker Model	Theoretical
Weidner (1995)	US air network	Airports	System dynamics	Theoretical
Helbing et al. (1997)	Human Trails	Origins, destinations	Active Walker Model	Theoretical
Schweitzer et al. (1998); Gastner and Newman (2006)	Idealized	Nodes	Minimization of cost function	Theoretical
Barabási and Albert (1999)	Idealized	Nodes	Preferential Attachment	Theoretical
Barabási (2002)	WWW, biochemical, communication, etc.	Nodes	Preferential Attachment	Empirical
Yamins et al. (2003)	Idealized	Land use lattice	Maximum potential	Theoretical
Levinson and Karamalaputi (2003a,b)	Twin Cities roads	Planning network	Heuristic expansion or construction rules	Statistical
Chen (2004); Levinson and Chen (2007)	Twin Cities roads and land use pattern	Planning network, land use lattice	Area-Based	Statistical
Levinson and Chen (2005)	Twin Cities roads and land use pattern	Planning network, land use lattice	Markov Chain	Statistical

Continued on next page

Table 2.1 – continued from previous page

Reference	Market	Given	Rules	Method
Xie (2005)	Idealized	Grids of different geometries	Network degeneration, "weakest link" heuristic	Theoretical
Yerra and Levinson (2005); Levinson and Yerra (2006)	Idealized	Grid	Agent-based system dynamics	Theoretical
Zhang (2005)	Idealized, Twin Cities	Planning network	System Dynamics	Theoretical
Mohammed et al. (2006*a*); Mohamed (2007)	Toronto bus network	Planning network	Regression, simultaneous equation	Statistical
Montes de Oca (2006)	Twin Cities roads and land use pattern	Planning network	Stated decision rules of jurisdictions	Empirical
Xie (2008)	Indiana interurbans	The full network	Network decline, "weakest link" heuristic	Empirical
Xie and Levinson (2009)	Idealized	Places, hexagonal potential network	Incremental connection	Theoretical

Chapter 3
Framework

This book aims to understand the complexity of transportation development and construct analytical explanations of the evolution of transportation networks. Transportation development is driven by the interaction of demand and supply, while subject to temporal and spatial constraints. Therefore, a meaningful model of network growth requires comprehensive consideration of the four dimensions of demand, supply, space, and time, which are briefly surveyed in the next four sections. Critical assumptions are also laid out, while their role may only become apparent in subsequent chapters. Figure 3.1 illustrates the mutual relationship between these key elements.

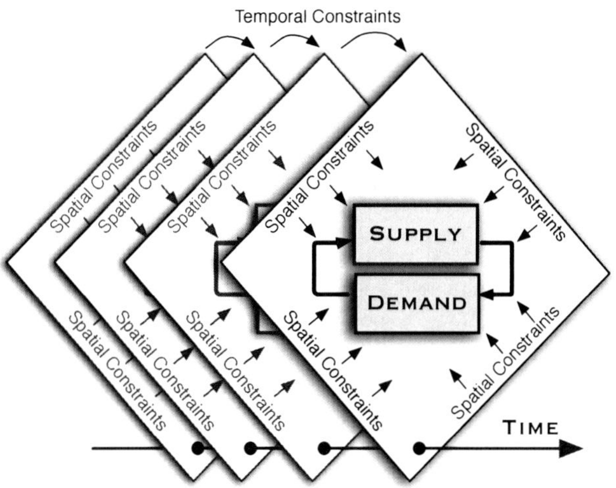

Fig. 3.1 A general framework of modeling transportation development

3.1 Supply

In this book, transportation infrastructure is represented as a geographical network where supply decisions are realized on individual links. Under this format, changes to a network only come from either expansion of existing links or construction of new links. During a specific time period, the discrepancy between the volume of traffic that traverses an existing link and the link's capacity to handle through traffic creates the need for additional capacity on the link. In terms of new construction, it is assumed that new links always connect to established places. However those links may create new places when they intersect.

Traditionally, change to a network is believed to be the outcome of a top-down policy-making process by which a central authority, in an attempt to maximize the overall efficiency of the system, provides an optimum amount of transportation supply subject to budgetary and other constraints.

From an evolutionary point of view, on the other hand, transportation supply is more than one-dimensional. Transportation infrastructure could be provided by either public or private suppliers under distinct interests. For example, while Chapters 4 to 6 look into historical transportation networks including skyways, interurbans, and streetcars that were in large part privately funded, Chapters 13-15 examine road networks under governmental provision. At different geographical scales, a network could be provided by a dominant supplier, such as a central government (e.g., the national Interstate system) or a monopoly enterprise (e.g., the Twin Cities streetcar system); or provided by a group of decentralized suppliers such as adjacent local jurisdictions (e.g. county roads) or competing private enterprises (e.g. Union Traction Company, one of more than 20 interurban companies of Indiana); or a mix of suppliers at both local and regional levels. Table 3.1 summarizes the diverse perspectives this book takes on the ownership organization of transportation systems.

It should be noted that there is an additional dimension that divides transportation networks into priced or unpriced systems. For instance, passengers traveling by interurban were charged a fare, but pedestrians using the skyways of Minneapolis are not.[1] Publicly provided infrastructure can be priced as well. In recent years, road pricing, particularly congestion pricing, has gained its popularity among both researchers and practitioners. With the exception of Chapter 14, this book does not bring particular details to pricing issues in spite of its importance, as its main interest is in transportation investment processes.

Another polychotomy of these studies is that chapters in Part II analyze the sequence or pattern of network changes based on historical observations, chapters in Part III and Part IV are theoretical explorations that demonstrate the transformation of transportation networks over space in abstract models of network evolution, while chapters in Part V undertake the provision of transportation networks from

---

[1] While developers of both network were local private enterprises, the latter system was built to make a profit from the increased value of connected businesses or land use, rather than from the use of network.

Table 3.1 Chapters that examine different ownership structures in transportation supply

		Ownership	
		Public	**Private**
Geographic scale	**Local**	13, 14, 15	4, 5, 8, 9, 12
	Regional	10, 11,13, 14,15	6

a normative approach, mainly examining the governance and forecasting issues in transportation supply.

Taken together, we can see that the supply of transportation infrastructure, rather than following an optimal design imposed by central authorities, is the outcome of the continuous interplay of travelers, business owners, suppliers, and operators under independent interests. Based on this rationale, agent-based simulation is adopted to model transportation development as an evolutionary process of interdependent decision making, which will be elaborated in the subsequent chapters.

3.2 Demand

The development of land use (demand) and transportation (supply) is a coupled process in which each drives the other. While its essential role in transportation development is fully recognized, transportation and land use interaction involves intricate modeling processes and merits a separate study in its own right. With a main focus on the transportation side of the story, this book generally treats land use as exogenous input to network models. In Part IV, however, land use is endogenously treated with network growth. Chapter 11 models place formation coupled with the diffusion of a transportation network over space, and Chapter 12 explicitly models the spontaneous organization of transportation and land use in a co-evolutionary process.

One could reasonably argue that transportation networks are created to serve the demand of moving people and freight to their desired destinations. In this regard, one would posit that a transportation network expands to where it is demanded the most. When facing a set of isolated places, the network is expected to first connect to most valuable destinations. In order to predict the sequence of link additions during network deployment, the value or attraction of place needs to be determined first. Accessibility, defined as the ease of reaching land use over a network, provides a measure of attraction impeded by travel cost, and is examined as a predictor of network extension in both empirical and simulation models of the book. In a network where places can be both travel origins and destinations, it is further posited that a network expands to where the overall accessibility of the system is maximized, while to which extent accessibility is aggregated depends on the ownership organization of the system.

There are exceptions, however, when induced demand gains the upper hand on induced supply. Networks are sometimes built speculatively based on anticipated future development. These networks, rather than responding to the demand, usually develop into open vacant land and induce new residences and businesses. Subsequent analysis will detail this phenomenon.

3.3 Time

Rome was not built in a day, nor are transportation networks. The ongoing growth of networks are usually measured in decades, if not in centuries. Over such a long duration, network growth is usually treated as a sequential process, by which changes to a network are implemented in discrete periods. This is consistent with the current practice of transportation engineering and planning, in which designs and plans are drawn in discrete planning horizons, land use / travel demand forecasting are carried out in incremental periods, and transportation projects are reviewed and funded periodically.

The rationales behind the discrete view of network growth are several. First of all, changes to transportation networks are the cumulative outcome of individual decisions made by professionals and policy-makers. Even if they are attempting to make optimal decisions, the "optimality" is constrained by their knowledge, information, budget, and other resources. From a long-term view, these decisions are at best "local optima" within a limited time horizon. In this regard, a discrete representation of transportation development better reflects the actual process by which supply decisions are undertaken subject to various constraints.

Second, what a network is today is restricted by what the network was in the past, which is widely referred to as the "legacy" phenomenon, a mutually reinforcing constraint structure, or obduracy in the terminology of Hommels (2005). The "legacy" effect has played a prominent role throughout the history of transportation development. London after the Great Fire of 1666 serves as a good example (Hanson, 2001). Since the fire destroyed most of the city including its street network, several radical rebuilding schemes were proposed, including a famous plan by Christopher Wren (Jardine, 2002). If any of them had been carried out, the city would have been significantly different. None of these schemes were actually realized, however, for the complexities associated with ownership of houses and lots (regardless of the fact that most houses were physically destroyed) and compensation in large-scale re-modeling, and eventually the old street network was re-built with only minor improvements in hygiene and fire safety. With the effect of historical legacy well recognized, it is not unreasonable to assume that supply decisions on a network are made in the current period treating the network from the previous period as given.

Last but not the least, sunk costs for infrastructure are high. This leads to the virtually irrevocable path of network growth that is notably sensitive to its initial conditions. Depending on the initial conditions, a certain sequence of network developments may lead to lock-in or market failure even if each individual decision has followed local optimization. A notable example is Beijing's current road system

(Huang, 2004). While the construction of a sequence of six ring roads demonstrates Beijing's determination to disperse its aggravating traffic and air conditions, the current situation may largely be attributed to the initial decision some fifty years ago to develop its urban transportation systems surrounding the historic city core. Taking this, a sequential realization of network growth is adopted in this book to better capture the path-dependent feature of transportation development.

3.4 Space

Instead of moving people and freight directly from one point to another, surface transportation infrastructure takes the form of a geographical network that is subject to spatial constraints. The network is hierarchical in nature as some of the links or nodes are more important than others. The notable development in network analysis, travel demand forecasting, and optimization algorithms since the 1960s have provided a variety of tools and techniques to predict disaggregate travel behaviors and demand across a network. This book adopts a general bi-level framework for the modeling of network growth, in which models at the lower level predict travel behaviors such as destination and route choices on a given network,[2] while the upper level models suppliers' decision-making processes based on the equilibrium results from the lower level, which in turn make change to the network and update its topology. The temporal growth of the network is realized by repeating this two-way process in consecutive periods.

Because of its vast investment, transportation infrastructure cannot be easily moved. Agencies build and operate links considering local factors, discouraging competition, and establishing a spatial (and generally a natural) monopoly. When a network is spatially connected but geographically divided into decentralized ownership, positive externalities may exist across local systems as travelers may use infrastructure they do not pay for. This is generally referred to as spatial spillover or the free rider effect. The conflicting forces of spatial monopoly and spillover effects are the key to understand the behaviors of local suppliers or operators in a decentralized network system, which will be extensively examined in subsequent analyses.

3.5 Summary and discussion

This chapter discusses the supply, demand, temporal, and spatial dimensions of transportation systems that are central to the growth of transportation networks. The increasing recognition of the evolutionary, interdependent, and path-dependent nature of transportation development in recent years has called for an innovative per-

---

[2] Despite its increasing popularity, multimodalism is not in the scope of this book. To simplify, traveler' mode choice is neglected in the models by assuming a single, abstract mode.

spective on transportation research and planning, which constitutes the main theme of this book. Following this theme, this book carries out a series of stand-alone studies which approach the question from different angles empirically, analytically, and in simulation.

Part II

NETWORK GROWTH IN THE PAST

This part presents empirical studies that explore the deployment of modern transportation systems such as skyways, interurbans, streetcars, rails, roads, aviation and seaport systems using *ex post* data.

Chapter 4
Skyways in Minneapolis

4.1 Introduction

Often enclosed and climate controlled, skyways link second level corridors across buildings, connecting various activity hubs such as shops and offices (Robertson, 1994; Byers, 1998). Such links allow for more efficient movement of pedestrians while protecting them from weather and the hazards of vehicular traffic below. Among North American cities that have substantial skyway systems, Minneapolis witnessed the city's first skyway link in 1962 as a modest effort to provide greater access to the central business district (CBD) (Byers, 1998). Over the next four decades, a system of skyway links has emerged, resulting in a network that connected over 70 continuous blocks in downtown Minneapolis. Corbett et al. (2009) comprehensively recapitulates the evolutionary history of the Minneapolis skyway system.

Minneapolis's first skyway bridge connected the Northstar Center, downtown's first mixed-use building, with the Northwestern National Bank across Marquette Avenue in 1962. The next year, a connection to the Roanoke Building over 7th Street was also completed. The original intent of these connections, made possible by collaboration between the Minneapolis City Planning department and local business leaders, was to make the financial district "more convenient for business people and clients to traverse" (Jacob, 1984).

In 1969, three skyways were added to the system. One skyway linking the Radisson Hotel to the Radisson Mart and parking was built over 7th Street. Another skyway spanned 6th Street between the FirstStar Bank (Northstar Center) and the Rand Tower, connecting four blocks of the financial district. The 5th skyway connected Dayton's Department Store with the LaSalle court over 8th Street. This was the first skyway to connect establishments in the retail core (Kaufman, 1985; Byers, 1998).

Perhaps the most significant early addition to the system occurred when the IDS Center opened its doors in 1973. The block-sized, mixed-use complex featured four skyways that connected adjacent blocks in each direction. More importantly, the skyway over Nicollet Mall connected the financial district with the retail core, thereby establishing itself as the center of the system. It quickly became a new landmark for the city and captured the imagination of city dwellers and additional business leaders regarding what skyways could do for downtown (Kaufman, 1985; Byers, 1998).

The construction of the early skyways drew much traffic and was deemed a tremendous success. Pedestrian traffic multiplied with the addition of more skyways, as did the number of merchants demanding space along these corridors. The property values of the second level rose, while values in the first-floor held up (Kaufman, 1985). As a result, the overall value of each connected building increased. By the early 1970s, retail space on the second level rented for as much as street-level space. Skyway connections were seen as an amenity to a building, thus owners could command higher lease rates for office space. As retail leasing rates steadily climbed throughout the 1980s, and in some cases were twice the value of some street level rates (Kaufman, 1985), owners and developers, fueled by the desire to add value to their properties, had new buildings designed and old ones retrofitted to include skyways.

By 1975, thirteen blocks within the retail and financial core of downtown were connected to the skyway system. From 1975 to 1985 the system grew at a rapid pace. Much of the development occurred at several sites on the fringes of the CBD (Byers, 1998). On the eastern edge of downtown, the Hennepin County Medical Center (HCMC) connected their buildings together, but was too far east to connect to the rest of the system. Substantial growth occurred in the Gateway District, just north of the CBD, linking office buildings and apartment high rises together. Additional skyways linked the Government Center to areas south and east including a municipal parking ramp, the Lutheran Brotherhood Building, and Centre Village. The new Piper Jaffray Building linked itself and the Energy Center to the main system near the financial core, and City Center, a shopping and entertainment complex completed in 1983, connected two additional blocks to the retail core.

During 1986-1995, the skyway system expanded along the South Mall to Orchestra Hall, and then onto the Convention Center. Significant skyway expansions were built to connect the Third Avenue parking ramps on the western edge of downtown with the retail and financial core of the system to entice auto travelers to patronize them (Byers, 1998). There were other skyways built, two of which provided better connectivity between the retail and financial cores, but most of the construction was sponsored by the City of Minneapolis government and accounted for fifteen blocks that were added to the system.

Between 1996 and 2004, the pace of skyway construction had slowed. A few blocks in the southwestern portion of the CBD were connected due to the new Target Headquarters and adjacent store and St. Thomas University expansion efforts. The completion of the Block-E retail/ entertainment center also occurred. Figures 4.1, 4.2, 4.3, 4.4, displaying the "snapshots" of the skyway system in 1975, 1985, 1995, and 2004, respectively, illustrate how the network has expanded in downtown Minneapolis over the past four decades.

As the system evolved, so did its governance. Unlike publicly owned skyway systems, such as those in Des Moines, Calgary, Cincinnati, and St. Paul, with their financing, construction, and management under direct city control (Montgomery and Bean, 1999), the Minneapolis skyway system is a privately supplied network through which private buildings are connected to one another. In the early system, all of the links were privately built and operated, thus operating hours depended upon decisions made by the building owners on each side of a skyway connection. This created problems, especially in the evening hours when some workers and shoppers were not able to return to their cars the same way they came in. There were also concerns about skyway bridge standards and security. Since the 1980s, the city of Minneapolis started to regulate much of the skyway expansion. In 1980, the Minneapolis Downtown Council created the Skyway Advisory Committee[1] to set guidelines for minimum bridge widths, heights, and spans, and encourage owners to adopt uniform operating hours. The SAC provides design reviews and approvals for changes and additions to the skyway system, and serves in an advisory capacity to the Minneapolis City Council. Another important task includes setting standards for skyway system signage and navigational aids (Jacob, 1984).

Despite the increasing role played by governmental interests throughout the growth and development of the system, the evolution of the skyway network did not always align with what the City of Minneapolis had laid out in several visions of a skyway system connecting most of the blocks downtown (Jacob, 1984; Kaufman, 1985), leading some to criticize the seemingly haphazard growth (Byers, 1998). This chapter attempts to determine if the growth of the system followed a predictable path. One might posit that the system expanded to where it was valued the most. This may be reflected in the assessed worth of the surrounding unconnected blocks or perhaps in the number of people (i.e. jobs). In this research the point accessibility of each block is calculated to assess the values of a place before and after skyway construction; a connect-choice analysis is then carried out to determine if the accessibility measure can be used to predict the growth of the skyway system in downtown Minneapolis.

4.2 Methodology

4.2.1 Accessibility analysis

Accessibility measures the relative ease of reaching valued destinations (Hanson, 1995). To determine if accessibility could be used as a predictor of network growth, the point accessibility of each block lying within and adjacent to the connected skyway system had to be calculated right before each time the system was expanded. In this research, point accessibility for each of the blocks in downtown Minneapolis is determined using the following mathematical relation:

---

[1] The Committee consisted of 17 members that owned or occupied properties connected by skyways, plus six non-voting members including representatives from the City Coordinator's Office, Department of Public Works, City Planning Department, City Attorney's Office, Minneapolis Community Development Agency (MCDA) and the Department of Inspections.

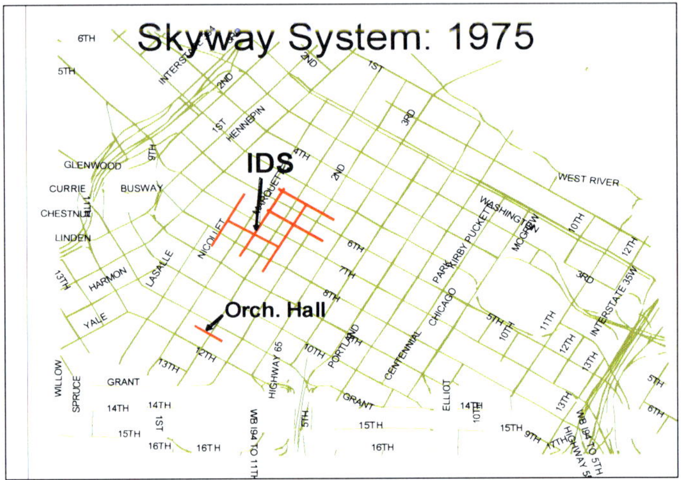

Fig. 4.1 The skyway network in downtown Minneapolis in 1975

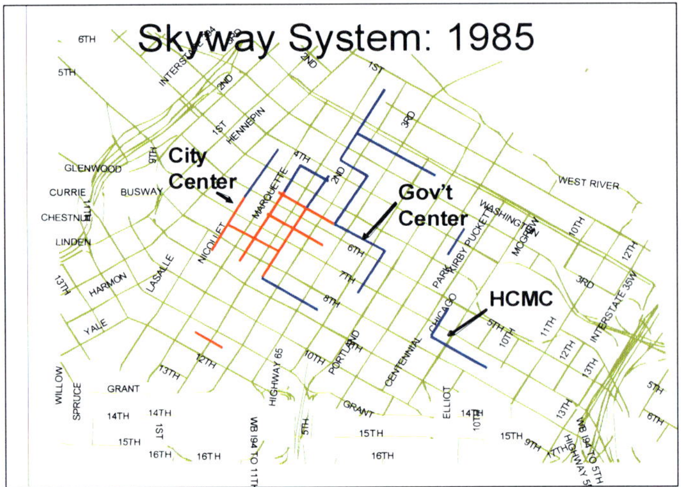

Fig. 4.2 The skyway network in downtown Minneapolis in 1985

$$A_i = P_i \sum_j P_j f(t_{ij}) \tag{4.1}$$

Where P_i and P_j represent some measure of activity, in this case, number of jobs held or number of trips generated in block i and block j, respectively; t_{ij} denotes the cost to travel between i and j along the shortest path, in this case, the least travel time walking from block i to block j; the cost function $f(t_{ij})$ is determined using a

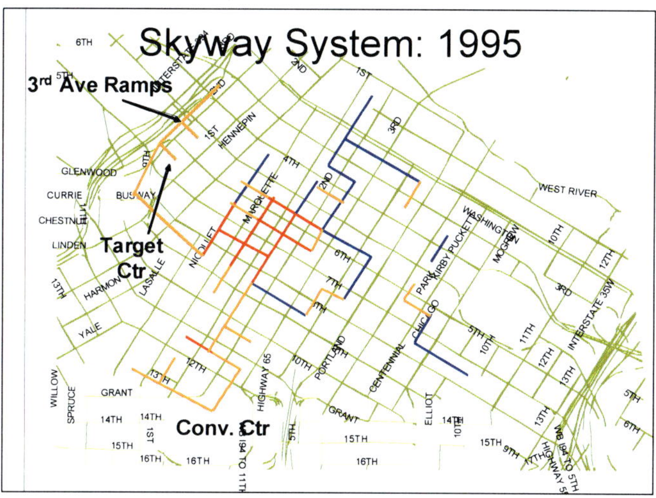

Fig. 4.3 The skyway network in downtown Minneapolis in 1995

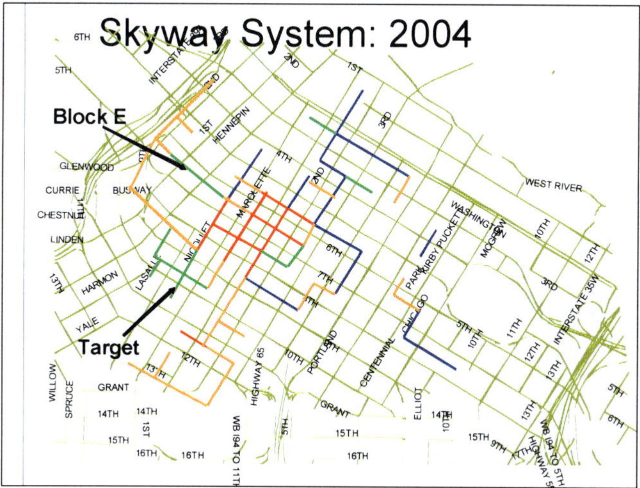

Fig. 4.4 The skyway network in downtown Minneapolis in 2004

gravity model. For simplicity, this analysis specifies a cost function in the following
quadratic form:

$$f(t_{ij}) = \frac{1}{t_{ij}^2} \tag{4.2}$$

Measurement of point accessibility required estimating the number of job opportunities or number of trips at each block and the travel times between each pair of

the locations in the skyway system. The number of employees on each block was determined from the square footage of the buildings located there. Current data on square footage was obtained from Kramer (2004) and the City of Minneapolis and Hennepin County's Property Finder/Information websites. Historic maps and other books dealing with the history of downtown Minneapolis were used to determine what buildings were in place for each of the years the skyway system experienced growth.[2]

Trip generation rates were obtained from the Institute of Transportation Engineers (1997). Since most of the land use in downtown Minneapolis was devoted to office and retail, determining trip generation rates for each block was not difficult. The ITE Trip Generation Handbook did not have parking ramps listed as a land use, so parking ramp occupancy data was obtained from the City of Minneapolis (which owns most of the ramps) to estimate trip generation rates. The website for the City of Minneapolis stated there were currently 161,000 people working in downtown's 3.6 million m^2 (39 million ft^2) of office and government space. Using these numbers, an average of 22.5 m^2 (242 ft^2) per person was calculated and subsequently used to determine the number of employees on each block connected by the skyway system. Whether or not this average held true back to 1962 is debatable, but due to a lack of data in this area it seemed to be the best option.

It needs to be noted that, although point accessibility has been calculated separately using two measures of land use activity, namely number of jobs and number of trips, and used as the input to the connect-choice model explained below, the model was found to perform better with point accessibility calculated based on trip rates. It was not surprising as the activities that took place on parking ramps would otherwise be ignored if only number of jobs in each block is considered. To save space, only results generated based on estimated trip rates are reported in the next sections.

Estimation of travel times between each pair of the blocks (all blocks are considered origins and destinations) required several steps. First, a street map of Downtown Minneapolis in GIS format was downloaded from the Metropolitan Council's DataFinder website. Using ArcMap, the skyways were then manually added over the street grid as shown in Figure 4.5. The network under examination consists of 118 street segments and 73 skyway connections. For simplicity, it was assumed that all of the skyway connections were straight paths from the mid-block of one building to another. It was also assumed that connections from one block to the next that did not have connecting skyway links (street only) were straight paths from the center of one block to the center of the other block. Similarly, links that had the

---

[2] Gathering data on historic structures proved difficult. The property information websites often did not yield data on structures that no longer exist, even if their addresses were provided. Old maps usually showed the footprints of buildings, not the number of square feet or stories. Old photographs/illustrations were sometimes useful for estimating the square footage of buildings, but it was often difficult to estimate how much of the block these structures covered. In the absence of good data, educated guesses were made. In the cases where data on particular blocks were not available, it was assumed the blocks contained a collection of 3-5 story buildings that covered approximately 75% of the block. This translates into 28,000 m^2 (300,000 ft^2) and is comparable to the current makeup of many of the blocks in Minneapolis's Warehouse District.

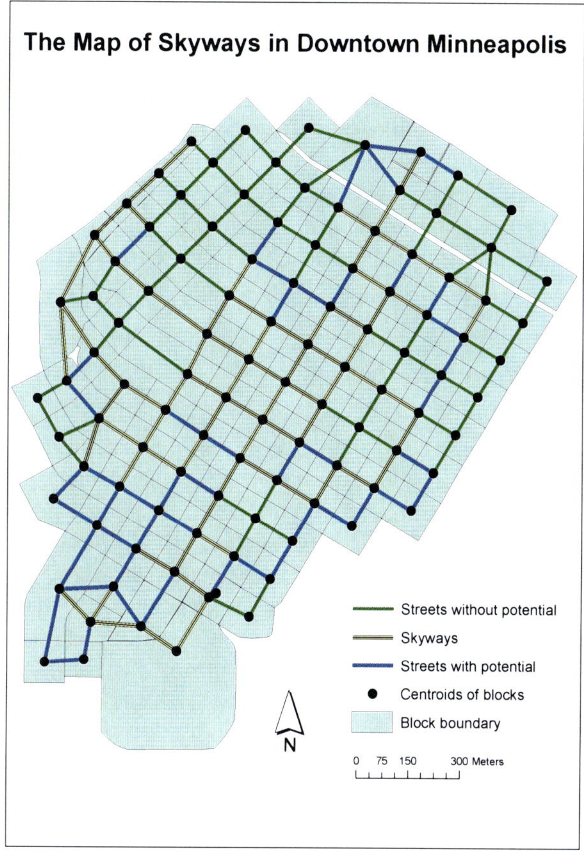

Fig. 4.5 The idealized skyway network in downtown Minneapolis

potential to be connected to the skyway system were added as links connecting the center of one block to the center of the next block. Next, nodes were created at the center of each linked block. The location of each node was determined from the ArcMap coordinates. The length of each link connecting one node to the next was then calculated according to node coordinates.[3]

Calculating the accessibility of each of the blocks involved an additional set of assumptions. The average walking speed was assumed to be 4.8 km/hr (3.0 mi/hr). However, in the case where travel had to be made from one block to another without using a skyway, the average walking speed was reduced to 2.4 km/hr (1.5 mi/hr).

---

[3] The map distance across a link does not necessarily represent the real distance one covers moving from one block to another. Between non-skyway blocks one has to go to the nearest intersection, and cross the street, thus increasing distance. The increased distance in the non-skyway case is accounted for by assuming a lower walking speed to cover the map distance of a link. See next paragraph for explanation.

This was done to account for the extra travel time and waiting time to cross streets (especially since most street crossing are made at the ends of blocks, rather than mid-block). Thus we essentially assumed the skyway link between two nodes has half the travel time cost of a non-skyway link. While to some extent these assumptions are arbitrary, they are unlikely to affect the accessibility rank of unconnected buildings, since all are treated equally.

Using these assumptions, point accessibilities were calculated for each block for each year the skyway system expanded. A connect-choice analysis was then conducted to determine if the most accessible blocks were in fact connected first.

4.2.2 Connect-choice analysis

In order to determine how often the expansion of the skyway network connected the blocks with higher accessibility, a connect-choice logit analysis was carried out relating the probability of an unconnected block's joining the network (in a given year) to its accessibility and current network size.

An algorithm was first developed to extract potential skyway connections in an iterative process, and evaluate the impact of building each candidate connection on accessibility. The flowchart of this algorithm is illustrated in Figure 4.6.

It is worth noting that the algorithm is generalizable to a variety of transportation networks. Its application to the Interurban network of Indiana will be discussed in the next chapter. In the case of the Minneapolis skyway system, the algorithm was run for 21 iterations, each of which represented a year when at least one skyway was built since 1962. Each iteration includes the following steps:

- Step 0: Find all candidate links for skyway construction given the network topology in the corresponding year of each iteration. All the links that have not been built as skyways by the year of examination and do not connect to any restricted blocks (i.e., historic districts and parks) are identified as candidates.
- Step 1: Change a candidate non-skyway link into a skyway link. The speed for non-skyway links, or street links is 2.4 km/hr (1.5 mi/hr), while the speed for skyway links is to 4.8 km/hr. After a candidate link is converted into a skyway link, the improved travel time between any pair of blocks in the network is recalculated using the shortest-path finding algorithm. The speed on this link is then restored to 2.4 km/hr (1.5 mi/hr) because the skyway construction is hypothetical.
- Step 2: Evaluate the accessibility impact. Based on the improved travel time upon hypothetically constructing a particular skyway link, the increase in accessibility for the two blocks that this link immediately connects to is calculated, as is the increase in accessibility for all the blocks in the network.
- Step 3: Implement the actual skyway connections. After evaluating all the candidate links one by one by repeating Step 1 and Step 2, the links where skyways were actually built in the corresponding year are updated with their improved travel speeds (and thus travel times) and labeled as "1" . The candidate links where no skyways were built in the corresponding year are labeled as "0".

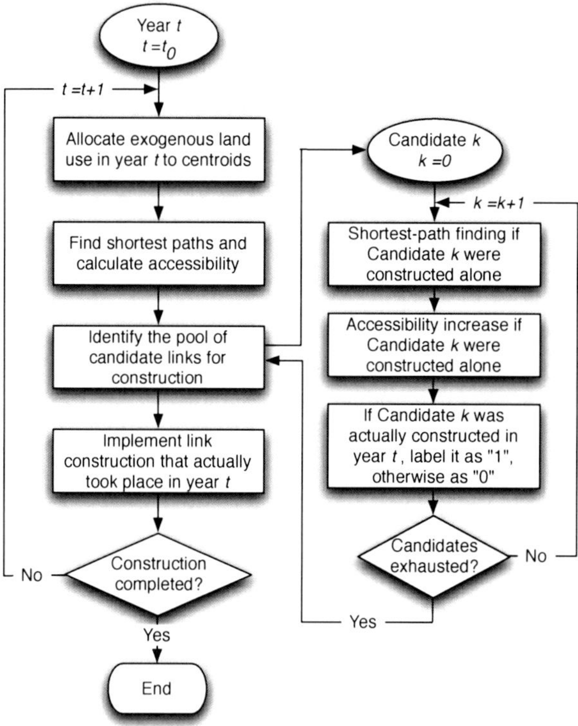

Fig. 4.6 Flowchart of the connect-choice analysis

Consequently, a dataset of 1,883 records was created, each representing a potential skyway connection for a particular year, with a binary variable ("0" or "1") indicating if this candidate connection was actually realized in this year. A logit model was then estimated using the statistical package STATA/IC 10.0 relating the binary variable to corresponding accessibility measures.

4.3 Results

As shown in Table 4.1, the connect-choice logit analysis indicates that the greater the increase in accessibility for the two connected blocks as a result of being connected to the network $(Ln(\Delta \tilde{A}_{i,j}))$,[4] the more likely the candidate link will be built.

---

[4] Variable $\Delta A_{i,j}$ denotes the potential change in accessibility for Block i and Block j if the skyway connection between the two blocks were added to the network. Because of the large magnitude of this accessibility measure (out of 1,883 observations of $\Delta A_{i,j}$, 102 report a value of zero, indicating little effect of skyway construction on accessibility, while the remaining 1,781 measures range from

This is suggested by the estimated coefficient for $Ln(\Delta\tilde{A}_{i,j})$ that is both positive and statistically significant. The influence of $Ln(\Delta\tilde{A}_{-i,-j})$ (the increase in accessibility of the remaining blocks), on the other hand, is not significant, suggesting that skyway links were not built with the goal of improving the overall accessibility of the region in mind. This result could be explained by the fact that most of the skyway links are built between private buildings and paid for privately by the owners of these buildings, who constructed skyways primarily to serve their own interests, that is, to increase the commercial activities in their buildings and to improve the related land value, while not caring of others' interests. As $\Delta\tilde{A}_{i,j}$ and $\Delta\tilde{A}_{-i,-j}$ demonstrate strong correlation (the correlation coefficient is 0.72), the product of the two variables $(\Delta\tilde{A}_{i,j} \cdot \Delta\tilde{A}_{-i,-j})$ is also included in the model to control their cross-effects. Other than the accessibility measures, we have included three explanatory variables: *year* (indicating the year of examination) and *size* (denoting the total length of constructed skyways in the year of examination) to eliminate possible time trend effect of connect-choice, although we realized *year* and *size* may be highly correlated; and $size^2$ (total length squared) to test if the connect-choice has a non-linear (quadratic) relationship with the total length of the network. As shown in Table 4.1, none of the coefficient estimates before the three variables turned out to be statistically significant, suggesting the inclusion of them did not significantly influence the probablity of skyway construction. Above all, though the overall predictive power (as suggested by the pseudo R-squared of 0.18) is relatively low, the logit model reveals that the increase in local accessibility ($\Delta A_{i,j}$) is a significant predictor of network growth.

4.4 Findings and concluding remarks

This chapter has demonstrated in the case of the Minneapolis skyway system that the expansion of an urban transportation system has to some extent followed a predictable path. A connect-choice logit model relating the probability of joining the network (in a given year) to accessibility corroborates the hypothesis that accessibility is a significant explanatory factor of network growth, although the estimated statistical model is not determinative. The connect-choice analysis further discloses that the greater the increase in accessibility for the two connected blocks as a result of being connected by a candidate link, the more likely the link will be built. The overall accessibility improvement in the rest of the blocks, on the other hand, is not a significant predictor. This finding is consistent with the private supply of the Min-

---

6.21E3 to 1.01E8, averaging 5.29E6), this model includes the natural logarithm of the variable as an explanatory factor. Taking an logarithm directly on zeros, however, would eliminate the corresponding records from the data set. To address this issue, a common practice in statistics is to add a small positive value to ensure the positiveness of a variable such that its logarithm always exists. In this case, one unit is added to each accessibility measure and the updated variable is denoted as $\Delta\tilde{A}_{i,j}$ (i.e., $\Delta\tilde{A}_{i,j}=\Delta A_{i,j}+1$). Considering the large scale of the accessibility measure, this transformation will not affect model results significantly.

Table 4.1 Connect-choice logit model results for the Minneapolis skyway system

Number of observations = 1883
LR $\chi^2(5)$ = 95.91
Prob > χ^2 = 0.00
Log likelihood = -224.65
Pseudo R^2 = 0.18

variable	coefficient	std. error	z	P-Val.	95% Conf. Interval	
$Ln(\Delta\tilde{A}_{i,j})$	0.73	0.19	3.94	0.00	0.37	1.09
$Ln(\Delta\tilde{A}_{-i,-j})$	0.019	0.17	0.11	0.91	-0.32	0.35
$\Delta\tilde{A}_{i,j} \cdot \Delta\tilde{A}_{-i,-j}$	1.92E-15	1.03E-15	1.87	0.062	-9.43E-17	3.94E-15
$size$	-9.02E-08	2.18E-04	-0.00	1.00	-4.28E-04	4.28E-04
$size^2$	-6.15E-09	7.85E-09	-0.78	0.43	-2.15E-08	9.24E-09
$year$	0.057	0.060	0.94	0.35	-0.062	0.17
$const$	-126.64	118.60	-1.07	0.29	-359.09	105.81

neapolis skyway system, and suggests that network ownership plays a pivotal role in the development of an urban transportation system.

The network expanded to the blocks with high accessibility with only a few exceptions. One reason why the most accessible blocks were not always connected first may have to do with some of the specific physical characteristics of the buildings (i.e. connection difficulties and lack of logical entry points into the buildings). A number of building owners may also have been averse to the idea of losing rentable office space. More savvy building owners in less accessible locations may have felt the potential benefits of being connected were speculatively substantial and pushed aggressively to be included in the system. Many of the skyways connected to parking ramps, which generally had relatively low accessibility values due to the fact that most of them are located on the edges of downtown. In addition, when the Skyway Advisory Committee came into play as a regulator of skyway development since 1980, politics and redevelopment objectives also played a role, especially in the 1990s as the city sought to connect the Convention Center to the system (presumably to attract more conventions and bring more convention goers into the retail and restaurant areas) and the Third Avenue parking ramps (to encourage more patronage). This study could be extended to include more sophisticated considerations of ownership, alliance, and the conflicts between private and public interests during the development of the skyway network.

Chapter 5
Interurbans in Indiana

5.1 Introduction

During the first decade of the 20th century the prevailing means of passenger transportation, steam-power trains, faced increasing challenges from electric railways and motor vehicles. Offering greater convenience and flexibility than the steam railroad, the electric railway had remarkable success in urban service for short-distance travel, which quickly led to its rural and intercity operation. The urban electric railway is referred to as the *streetcar*, while in rural use it is called the *interurban*. The two systems differ in that the equipment of the interurban is usually larger, heavier, and faster. While the next chapter will discuss of the streetcar, this chapter is directed to the interurban.

Among those surface transportation modes that have thrived and then declined in North America (including canals, turnpikes, steam-power railroad train, streetcars, and interurbans), the interurban probably experienced the most dramatic change. Most interurban rails were built between 1901 and 1908, and by 1912 the interurban network had taken its final shape 25,000 km (15,500 miles) in the US. A marked decline set in about 1918, largely due to the competition from the automobile with even greater flexibility, and within two decades the network was virtually annihilated. As the interurbans experienced such a short and relatively recent life cycle, the history of these interurbans is well documented and retrievable. Hilton and Due (1960) presents a comprehensive review of this industry.

Over all phases of its operations, the interurban remained independent enterprises, promoted, owned, and operated by private interests.[1] Some of the independent lines were financed by wealthy local businessmen; a number were promoted by syndicates formed by large urban financial groups, and usually developed into independent interurban systems of substantial magnitude through continuous exten-

---

[1] As Hilton and Due (1960) pointed out, public control over the interurban during the early years was limited to franchise requirements imposed by local government, such as paving requirements and restrictions on fares; state and federal control over the industry did not make a notable presence until 1907.

sion, new construction or acquisition across a region. Despite their occasional co-operation in developing joint ticketing and time tables, these interurban companies retained natural monopolies for local passenger transportation but were competitors for longer intercity trips.

Reaching a maximum of 2,937 km (1,825 miles), Indiana was second only to Ohio in the absolute size of its interurban network, and the only state where a large-scale grid-like interurban network emerged. Therefore the Indiana network has been the subject of interest to researchers for a long time (Haley, 1936; Marlette, 1959). The network had its first line in 1887, started to decline from 1917, and completely disappeared in 1941. In its complete shape, the topological pattern of the network can be best described as a series of irregular wheels, with their spokes radiating from major cities such as Indianapolis and Fort Wayne. By 1910, more than 20 companies/syndicates had been chartered in Indiana to build and operate interurban lines. For example, Union Traction Company, the largest interurban company in Indiana, had 660 km (410 miles) of interurban lines and 17 routes in operation; the Terre Haute Indianapolis and Eastern Traction Company, the second largest, operated 647 km (402 miles) of lines and 15 routes.

This chapter, extending the previous chapter's investigation into the Minneapolis skyway system, examines the growth of the Indiana interurban network during 1887-1916. For the purpose of this study, the network of Indiana is separated from those other states basically along the state border. The portion in north Indiana is excluded because it was more connected to Chicago and the Michigan cities around Lake Michigan, than to the main body of the Indiana network. For simplicity, detour lines shorter than 3.2 km (2 miles) are neglected and the locations of the places as the terminals of these lines are adjusted as if they were located on the main line, including Milton, Richmond, Gas City, and Riverside Park. Consequently, a network of 53 places and 62 interurban rail segments is extracted for examination as shown in Figure 5.1, which represents the interurban network of Indiana in 1916 in its "full" shape. Hilton and Due (1960) recorded the open/close dates of each interurban line in Indiana so that the actual topological evolution of the network can be retrieved through years. The franchise owner of each line is also displayed in Figure 5.1. Table 5.1 lists main franchise interurban companies in Indiana and their abbreviated names according to Hilton and Due (1960). This research aims to answer several questions including whether the deployment of interurban lines in Indiana had followed a logical path, whether accessibility, as in the Minneapolis skyway system, predicted where the network expanded, and whether different ownership organizations of the two systems had resulted in different courses of link additions.

The remainder of this chapter proceeds as follows: the next section introduces connect-choice analysis, which is followed by two alternative hypotheses of ownership organization of the interurban system; then results are discussed and conclusions drawn.

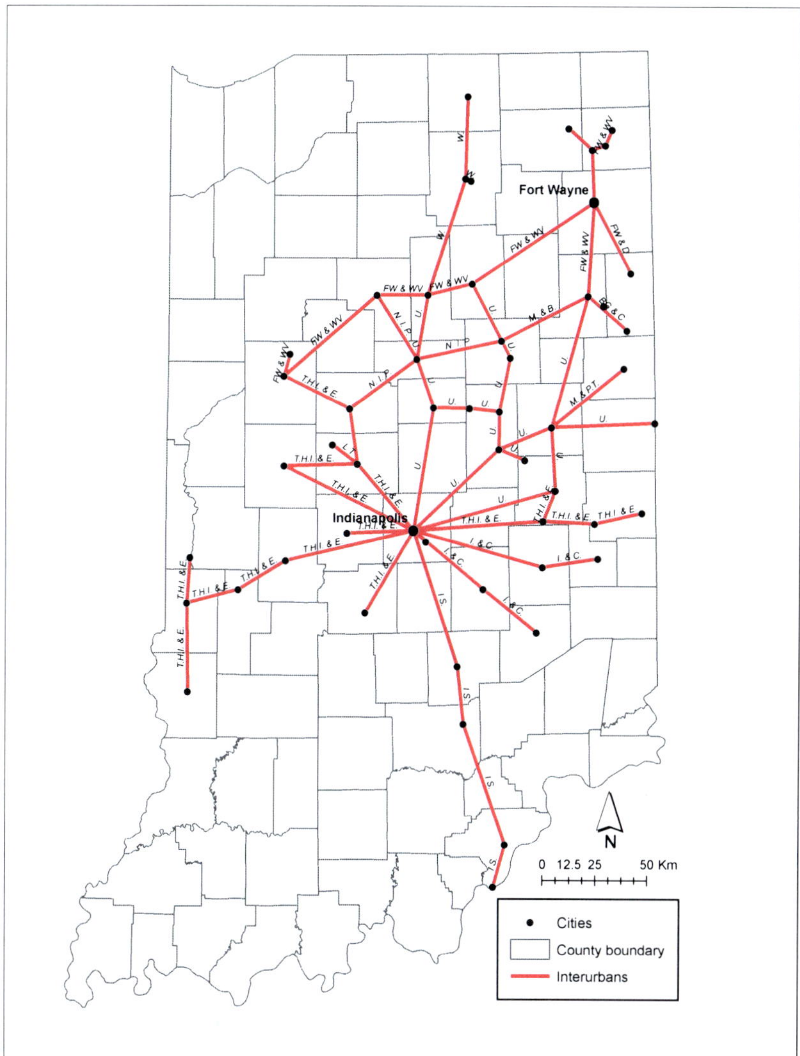

Fig. 5.1 The extracted interurban network of Indiana in 1916

5.2 Connect-choice analysis

In order to determine if accessibility is an effective predictor of line additions during
the deployment of Indiana interurbans, a connect-choice analysis is conducted for
the Interurban network during the period of 1887-1916. To be consistent with the
connect-choice analysis of the skyway system presented in Chapter 4, the general
procedure displayed in Figure 4.6 is followed, and specifics for this analysis are
explained below.

Table 5.1 List of franchise interurban companies in Indiana

Company	Abbreviation
Bluffton Geneva & Celina Trac. Co.	B.G. & C.
Northern Indiana Power Co.	N. I. P.
Fort Wayne & Wabash Valley Trac. Co.	F.W. & W.V.
Marion & Bluffton Trac. Co.	M. & B.
Fort Wayne & Decatur Trac. Co.	F.W. & D.
Indianapolis & Cincinnati Trac. Co.	I. & C.
Interstate Public Service Co.	I. S.
Lebanon - Thorntown Trac. Co.	L. T.
Terre Haute Indianapolis & Eastern Trac. Co.	T.H.I. & E.
Union Trac. Co. of Indiana	U.
Muncie & Portland Trac. Co.	M. & P.T.
Winona Interurban Ry.	W.

First, census population data by county are taken as exogenous input to the analysis. The state of Indiana consists of 92 counties, and the historical population records of each county are available every decade from 1890 (US Department of Commerce Bureau of the Census, 1996). Population of each county for each year is estimated by linear interpolation. Population needs to be allocated to stations[2] to conduct the subsequent accessibility analysis. Assuming the population of a county is allocated to the nearest station/city, determined using the airline distance from the centroid of this county to the station, 10 out of 53 stations would be assigned no passengers, which is obviously unrealistic. To resolve this issue, the census tract is instead used as the geographical unit to hold population information. With no historical population data by census tracts available during the analysis period, however, the population of a county is assumed to be evenly distributed and allocated to tracts within its boundary in proportion to their areas. Considering that most people have to access the stations on foot, horse, carriage, or electric streetcar in cities at that time (auto ownership was not yet widespread, and auto drivers would be less likely to use an interurban), it is also assumed that only people living in the tracts within 32 km (20 miles) of stations would consider taking interurban rail.

As an estimated 94.4 percent of the revenues of electric lines came from passenger fares (versus 5.6% from freight) in 1902 (Hilton and Due, 1960), this study includes only the change in accessibility to population as a predictor in the subsequent connect-choice analysis. Accessibility is calculated in a gravity-type model adopting Equations 4.1 and 4.2, where P_i and P_j in this case represent the population served by stations of city i and city j, respectively, while t_{ij} represents the least time an average person spends to travel from city i to city j by interurban. As no historical evidence shows that the interurban had a notable speed change over space, the

---

[2] For the sake of simplicity, it is assumed each city has only one interurban station where trains enter and exit this city. Considering the geographical scale of our analysis, though, this simplification is not expected to significantly influence the results.

interurban speed is assumed to be fixed over time and uniform across lines. Specifically, it is assumed that interurban lines have a uniform generalized speed of 24 km/hr (15 mi/hr) (taking into consideration a uniform fare per mile) while people can access the nearest station at an average speed of 8 km/hr (5 mi/hr).

Similar to the analysis conducted for the skyway network, for each year when at least one interurban line was actually built, potential line extensions to the existing interurban network are identified, and the benefit of each potential line is evaluated in terms of the increase in accessibility as the result of the proposed line being hypothetically constructed and added to the existing network. If a potential line was actually built in the year of examination, it is labeled as "1"; otherwise it is labeled as "0". At the end of each construction year the network is updated, and the process is iterated until the network reaches its final shape in 1916.

5.3 Hypotheses

One may posit that interurban lines, like skyways between private buildings of downtown Minneapolis, were constructed to primarily serve local interests. This leads to our first hypothesis that a proposed line that benefits local residents most has the highest possibility of being built. Based on the assumption that local promoters of interurbans always attempt to maximize the accessibility of local residents,[3] benefits associated with a line extension were evaluated as the increase in accessibility resulted from this line being hypothetically constructed. Under the further assumption that a line is constructed to maximize the accessibility of residents in the two cities connected by this line, changes in accessibility were then estimated for residents in these two cities and in all other cities, respectively. It is posited that that the increased accessibility for the former user group indicates a higher possibility of the proposed line being constructed, while that for the latter user group may be irrelevant to, or even counteracting the possibility of line construction. Under this hypothesis, a discrete connect-choice analysis was carried out to predict the likelihood of a potential line (connecting city i and city j for instance) being constructed in a particular year. Two explanatory variables ($\Delta A_{i,j}$ and $\Delta A_{-i,-j}$) were included to represent the increases in accessibility for residents of city i and city j, and that for residents of all other cities, respectively. Similar to the skyway analysis, a natural logarithm is taken on accessibility measures after one unit is added to each mea-

---

[3] Private entrepreneurs such as wealthy land developers were the primary promoters of local interurban lines. They had strong incentive to make the interurban more accessible and attractive, especially during the early deployment stage, in order to compete with other modes in the market. For those promoters who are closely associated with local real estate businesses, their wealth accrues from increased accessibility to their holdings, which provides another incentive for them to improve the accessibility of interurban. Local governments may choose to subsidize the interurban as well, not only because they are interested in improving their constituents' welfare associated with increased accessibility to jobs and other activities by promoting public transportation, but also because of the pressure from residents who demanded access to the interurban service in the exceptionally enthusiastic atmosphere towards this technology in that era.

sure to ensure positive values. The transformed variables are denoted as $Ln(\Delta \tilde{A}_{i,j})$ and $Ln(\Delta \tilde{A}_{-i,-j})$, respectively. It should be noted that $\Delta \tilde{A}_{i,j}$ and $\Delta \tilde{A}_{i,j}$ bear no correlation (the correlation coefficient equals -0.027), so it is not necessary to control their cross effect. Like the skyway analysis, a *year* variable was also included in the model to control the time effect of interurban construction.

The hypothesis of local provision may be plausible for some local lines deployed during the early stage, but it does not account for the initiatives of the syndicates that promoted the vast part of the network. Most of the syndicates owned and managed independent interurban systems on a regional level that served multiple cities with a series of routes, and competed with each other for intercity demand. Considering the oligopolistic nature of the syndicates that promoted the interurban system of Indiana, can accessibility still be used as an predictor of line extensions? To answer this question, it is natural to posit that a company constructs an interurban line not only looking at the local market, but also looking at the regional market served by its regional network. Thus an alternative analysis was proposed to associate the likelihood of a potential line being constructed with the increase in accessibility for the residents served by the syndicate (regional interurban company) that had acquired the franchise of this line (represented by ΔA_c, where c denotes the franchise company) and the increase for those not served by the syndicate (ΔA_{-c}). We posit that the higher benefit a potential line can bring to its franchise company (in terms of the increase in accessibility for the residents served by this company), the larger chance this line will get built. On the other hand, it is posited that a syndicate will avoid building a line that would bring more benefits to its competitors. In other words, the higher benefit a potential line will bring to competitor companies as opposed to its franchise company, the smaller chance it will get built. This hypothesis is proposed based on the fact that interurban companies at that time were all private enterprises and competitors of each other, although their networks were geographically interconnected.

5.4 Results

Binary connect-choice logit models under alternative hypotheses were estimated using STATA/IC 10.0, and their results are summarized in Table 5.2 and Table 5.3, respectively. While both including accessibility as an explanatory factor of line extensions, the two models develop different pairs of accessibility measures under different assumptions of ownership organization of the system. The first model includes the change in accessibility for residents in city i and city j ($Ln(\Delta \tilde{A}_{i,j})$) and that for residents in all other cities ($Ln(\Delta \tilde{A}_{-i,-j})$). As shown in Table 5.2, neither variable turned out to be statistically significant in the model, which does not support our first hypothesis that the network is developed primarily from local interests. On the other hand, Table 5.3 shows that changes in accessibility well predict line extensions in the consideration of syndicates' initiatives. As the result of a proposed line being hypothetically added to the network, a higher increase in accessibility for residents served by the franchise company predicts a higher possibility of this line being actually selected for construction (as suggested by the positive and significant coefficient before $Ln(\Delta \tilde{A}_c)$); in contrast, an increase in accessibility for residents not served by the franchise company indicates a smaller possibility of the proposed line being actually constructed (suggested by the negative and significant coefficient before $Ln(\Delta \tilde{A}_{-c})$).

Table 5.2 Connect-choice logit model results for the Indiana interurban network under the hypothesis of local provision of infrastructure

Number of observations = 466
LR $\chi^2(5)$ = 69.30
Prob > χ^2 = 0.00
Log likelihood = -146.20
Pseudo R^2 = 0.19

variable	coefficient	std. error	z	P-Val.	95% Conf. Interval	
$Ln(\Delta \tilde{A}_{i,j})$	0.18	0.13	1.42	0.16	-0.070	0.43
$Ln(\Delta \tilde{A}_{-i,-j})$	0.10	0.14	0.73	0.46	-0.17	0.37
year	0.35	0.056	6.21	0.00	0.24	0.46
const	-671.66	105.83	-6.35	0.00	-879.09	-464.23

Table 5.3 Connect-choice logit model results for the Indiana interurban network under the hypothesis of oligopolistic provision of infrastructure

Number of observations = 466
LR $\chi^2(5)$ = 251.32
Prob > χ^2 = 0.00
Log likelihood = -55.19
Pseudo R^2 = 0.69

variable	coefficient	std. error	z	P-Val.	95% Conf. Interval	
$Ln(\Delta \tilde{A}_c)$	1.51E-10	2.73E-11	5.53	0.00	9.75E-11	2.04E-10
$Ln(\Delta \tilde{A}_{-c})$	-3.14E-10	4.85E-11	-6.47	0.00	-4.09E-10	-2.19E-10
year	0.21	0.083	2.57	0.01	0.05	0.38
const	-407.89	158.38	-2.58	0.01	-718.30	-97.47

5.5 Findings and concluding remarks

This study retrieved the historical data from the interurban network of Indiana during its growth phase from 1887 to 1916, and carried out a discrete connect-choice analysis on the data to determine if accessibility is an effective predictor of link additions when the network expands. Analysis was carried out under two alternatives hypotheses with different assumptions on the incentives of owners of the transit system. The first hypothesis assumes local players, including local businessmen and governments in the two cities connected by a proposed line, are the primary promoters of this line, thereby calculating accessibility for residents of the two cities and for those in all other cities, respectively. A subsequent connect-choice analysis indicates neither of these two accessibility measures are significantly associated with the connect-choice. The second hypothesis, on the other hand, assumes syndicates who built and managed independent interurban systems and competed with each other on a regional level are the primary promoters of the system. An alternative connect-choice analysis accordingly included increases in accessibility for residents served by the franchise company of a potential line (as the result of this line being hypothetically added to the existing network), and for those not served by the company as explanatory variables of line additions. Results suggest that the former increase in accessibility predicts a higher possibility of this line being actually selected for construction, while the latter indicates a smaller possibility of line construction. This analysis not only provides empirical evidence that accessibility is an effective predictor of line extensions during the growth phase of the Indiana interurban network, but also discloses how interurban lines were deployed to serve the syndicates' own interests while against their competitors' in an oligopolistic organization of the system. While a syndicate tends to build a line to maximize the benefits of potential customers of its service, it also avoids building a line that could benefit its competitors.

It is worth comparing these findings with those of the Minneapolis skyway network. Both analyses corroborate our claim that the expansion of a transportation network follows a logical path by which the accessibility of the system is maximized. They further reveal that the two systems have followed different evolutionary paths under different ownership organizations. The skyway network was developed between buildings by local private owners of these buildings. The interurban network, on the other hand, was developed primarily by syndicates that owned and managed independent systems (networks) of interurbans at a regional level. While both are private systems, the skyway system is not priced as its developers are primarily interested in the increased value of surrounding businesses and land use; the interurban, on the other hand, charges passengers for the use of the network. It is shown in both studies, that our mathematical representation of accessibility is effective not only in the sense it was significantly associated with the possibility of link additions in both networks, but also in that it was able to account for the difference in ownership organization of the two systems by identifying different beneficiary groups from link additions. The accessibility analysis could find immediate application in preliminary transportation planning and network design considering their long-term effect on a system. This research also has implications on how planning efforts should be directed in the provision of transportation infrastructure to reconcile the interests of local versus regional, and public versus private promoters of a system taking into account the mixed ownership structure.

Chapter 6
Streetcars in the Twin Cities

6.1 Introduction

The story of every urban transportation system has been a product of the geography and demographics of its surrounding region, while the development of those transportation systems have in turn transformed the cities and communities they serve. In the 1960s, geographers recognized the marked role of land use-transportation interactions in the formation and differentiation of places and transportation networks (Haggett and Chorley, 1969; Lowe and Moryadas, 1975). Since the 1980s, integrated land use-transportation modeling has gained momentum and seen widespread application in urban planning and design.[1] In empirical research, the relationship between transportation and land use was widely examined as a two-way process by which one drives the other. A fraction of this literature focused on *Granger causality*[2] between land use and transportation development. Using county level data in the US Mid-Atlantic Region, Fulton et al. (2000) estimated cross-sectional time-series models that relate utilization (measured in distance traveled) to roadway capacity. The results indicated that growth in capacity preceded growth in utilization. Cervero and Hansen (2002), presenting simultaneous models that predict induced usage and capacity on roads using 22 years of observations for 34 California urban counties, found strong reciprocal relationships between road investment and travel demand. King (2011) explored the co-development of the subway system and residential and commercial land uses in the early 20th Century New York City using Granger causality models, finding that New York's subway network developed in an orderly fashion and grew densest in areas where development had already occurred, while lagged station densities were a weak predictor of residential and commercial densities. As evident from these studies, the reciprocal relationship between trans-

---

[1] Refer to Timmermans (2003) or Iacono et al. (2008) for comprehensive surveys of this literature.

[2] First introducded by Granger (1969), Granger causality allows us to determine whether changes in land use provide statistically significant information about future changes to transportation infrastructure, or vice versa.

portation and land use is pervasive, but induced supply and induced demand may exhibit different strengths in different transportation systems.

The reasons for the difference could be geographical, technological, economic, managerial, social or political. Levinson (2008a), analyzing the co-development of rail and population in London since 1841, found that the feedback effects between population density and rail density are distinct in the core and periphery of the city. While a positive reciprocal relationship between rail and population was found in the periphery, the causation effect was only one-way from rails to population in the city center (the accessibility created by rail allowed firms to displace residents in the core).

Chapter 4 disclosed that the addition of skyway connections in downtown Minneapolis had followed a mathematical path by which the accessibility to the activities in the connected buildings (measured by built area) would be maximized. Skyway deployment has largely followed rather than led downtown establishments probably because it is a lot less costly to add a skyway than to construct new space in the dense city center.

Similarly, the interurban railways in North America were deployed in large part connecting the established cities and towns (Hilton and Due, 1960). The coupled population growth in cities connected by interurbans was not as significant probably because the system was abandoned so quickly.[3] Warner (2004), using Boston 1870-1900 as a case study, described how the streetcar had led the process of suburbanization in American cities during the Gilded Age.

The streetcar system in the Twin Cities of Minneapolis and Saint Paul provides another example. Over the period from the 1870s to the 1950s, the system had been a significant force in the growth and shaping of the Twin Cities. Starting from two primitive horsecar lines in 1875, it had developed into one of the world's finest urban transportation systems over the course of the twentieth century's opening decades. In the 1910s, the electrified streetcars provided nearly 100 percent of all public transportation in the Twin Cities. At its peak, the system carried over 200 million passengers each year between 1919 and 1925 and reached its greatest extent, 842 km (523 miles) of track, in 1931. Diers and Isaacs (2006) recount the streetcar era in the Twin Cities.

As Adams and VanDrasek (1993) pointed out in a brief recap of the Twin Cities streetcar history, population had first led, then followed the deployment of streetcar lines. Born in the 1840s, Minneapolis and Saint Paul had emerged on the frontier as important gateway cities and rail centers for people that moved west following the railroads. As the city grew, urban population expanded rapidly. The Federal Census reported 297,894 inhabitants for Minneapolis and Saint Paul combined in 1890, compared with 615,280 in 1920, more than doubling over three decades. It became obvious that there had to be a faster and better way to move people than on foot or horseback. Following this demand, the first two horsecar lines were brought to Minneapolis in 1875. Interestingly, the two most notable promoters of streetcars

---

[3] The Indiana interurban system, for example, was built mainly between 1901 and 1908; the network started to decline in about 1918, and was completely abandoned within two decades.

in the Twin Cities, Colonel King and Thomas Lowry[4] were both actively involved in the real estate business. Colonel King had his farm and properties well outside the Minneapolis city limits, while Thomas Lowry developed many neighborhoods in Minneapolis, Saint Paul, and the surrounding communities. So it came as no surprise that they first introduced the streetcar in order to serve their outlying holdings and enhance their real estate value. As the city grew and people kept demanding public transportation services, they then built more streetcar lines at the interest of making more profits from the transit system. The system saw remarkable expansion from the initial 3.4 km (2.1 miles) of horsecar line in 1875 to 106 km (66 miles) by 1889.[5]

The first electric streetcar line opened in Minneapolis on December 24, 1889. Faster and cheaper than horsecars, the technology of electric railway proved to be a great success. By the close of 1891, nearly all the streetcar lines had been converted to electric power. As the streetcar system grew, so did the Twin Cities. The extension of streetcar lines opened up vast land and parcels in the suburbs, leading to the expansion of city boundaries. Thomas Lowry sold lots and homes at the terminals and intersections where residences and businesses subsequently sprouted, and even built amusement parks at the end of streetcar lines to encourage recreational trips, making enormous profits from both enterprises. The streetcar system was so prosperous and profitable that Twin Cities Rapid Transit (TCRT) eventually overbuilt the network. In fact, the rapid extension of streetcar lines was largely responsible for some of the lowest-density development of residential area found among midwestern cities (Adams and VanDrasek, 1993).

This chapter presents a causality analysis of streetcar deployment versus residential development in the electric streetcar era. Questions under scrutiny include whether streetcar extension and population growth exhibit causation effects with equal strength, and why one effect could gain the upper hand on the other. Based on empirical observations from the Twin Cities, this research aims to present empirical evidence on the significant role that the electrification of street railway has played in fostering urban growth and shaping residential patterns in the metro area. This chapter proceeds as follows: the next section presents hypotheses to be tested; data are then constructed and models proposed, which are followed by a discussion of results; in the last section conclusions are drawn and their implications highlighted.

6.2 Hypotheses

The development of the Twin Cities streetcar system demonstrated distinct characteristics in two different stages. The turning point was 1889, when the transit lines were changed to electric power. Prior to 1889, horsecar lines were deployed following a rapidly growing population; after 1889 the flourishing system extended into

---

[4] Thomas Lowry was the first president of Twin Cities Rapid Transit (TCRT), the monopoly operator of the Twin Cities streetcar system, and Colonel King was the initial organizer of Minneapolis Street Railway Company, the precursor of TCRT.

[5] Before 1889 steam cars and steam-power cable cars were experimented with, but quickly failed.

the open suburban land, leading to new residential development where streetcars reached. While most electric streetcar lines survived until the 1950s before being converted to bus routes, our analysis focuses only on the growth phase of the infrastructure, namely, 1889-1931. In review of the history, the rapid expansion of the Twin Cities streetcar system during this period had mainly been driven by three forces.

First, electric street railway technology was one of the most important inventions of the nineteenth century. As Diers and Isaacs (2006) stated, "few events in Twin Cities history in the nineteenth century can compete in significance with the electrification of the street railway system and its subsequent effects on growth and development". Compared with horsecars and steam-propelled streetcars, electric streetcars provide a faster and far more economical way to move people, while significant competition from internal combustion engine did not emerge until the 1910s and was not in full force until the 1920s. Similar to the British railway mania of the 1840s (Odlyzko, 2010), the superiority of the electric street railway technology at that time had resulted in the pervasive optimism among entrepreneurs,[6] and eventually led to the excessive expansion of the system.

Second, ownership structure played an important role in the deployment of the Twin Cities streetcar system. Unlike today's highways and transit systems, the streetcar system remained privately owned by a group of real estate investors and promoters. Their interest naturally led to the development of vast tracts of land where streetcar lines extended. When residential, commercial, and recreational activities followed the lines, both streetcar ridership and real estate value increased. On the other hand, unlike in other cities such as Philadelphia and Chicago where independent streetcar companies were granted franchises and eventually consolidated, TCRC was regulated as a monopoly since its early stage of development and throughout the electric streetcar era. The exclusive franchise, though, was granted by the city councils subject to two conditions: the first required a flat fare of five cents, and the second was the councils' power to order the company to build any line which the councils declared reasonably necessary (Lowry, 1978). For many years, these requirements assured an affordable ride on an efficient system, which enabled working people to live outside the city centers and commute to work.

Third, as then prevailing urban transportation, the streetcar system provided perhaps the only means of public transportation for the working class before the 1930s. During this era, proximity to a streetcar line determined where people lived and where they worked. Being accessible to a streetcar is so crucial that the deployment of streetcar lines had produced "finger-shaped extensions of residential areas" (Adams and VanDrasek, 1993) because "it was easy to build outward along the lines but hard to expand between them".

Based on the discussions above, it is posited that the extension of electric streetcar lines preceded the increase of residential density in neighborhoods reached by the streetcar. While we may also observe the causation effect in the opposite direction, that is, addition of streetcar lines in response to the increased demand (resi-

---

[6] As Diers and Isaacs (2006) described, the streetcar and the interurban were "dot-coms" of that era.

dences), we would expect the causation effect from the streetcar side to development is the stronger effect here. Rather than the existing demand, technological superiority, monopoly, close connections with real estate, and reliance on streetcars could be the major forces that drove the rapid expansion of the streetcar system.

Due to heavy reliance on the streetcar service and lack of complimentary transportation that would feed the streetcar, living close to streetcar lines was crucial for mobility to work and shopping at that time. We thus hypothesize that proximity to a streetcar line is the pivotal factor that determines the spatial distribution of residences. We expect to observe residential density declining outward along streetcar lines and significantly dropping beyond an affordable walking distance. It is also hypothesized that a neighborhood that gains closer access to the streetcar service (after new lines are opened) will see subsequent increases in residential density.

6.3 Historic Data

6.3.1 Network data

Time series population and streetcar network data are essential to test our hypotheses. Referencing the book by Isaacs and Diers (2007), Metropolitan Council, the regional planning agency serving the Twin Cities seven-county metropolitan area, has digitized in an ArcGIS shapefile all the historic transit routes in the region. The shapes contain streetcar, horsecar, ferry, 1948 bus and steam power routes, from which all the streetcar lines were extracted for our analysis. In the data set, one shape and one corresponding record represents each route segment (Metropolitan Council, 2007). Attributes of each record include opening year, closing year, corridor name, etc. In the data there are 12 streetcar route segments whose opening dates are missing. We corrected the absent opening dates referring to Olson (1976), which includes a comprehensive review of construction and abandonment for each streetcar line in the Twin Cities.

6.3.2 Population data

Analysis of the relationship between population and streetcars requires historic population data during 1889-1931 at a sub-county level, which turned out to be a challenging task. Our first resort was census population statistics at the tract level. Since census tracts were not laid out in the Cities of Minneapolis and Saint Paul until 1934 (Green and Truesdell, 1937), however, census tract data is not available for the electric streetcar era. Another possibility is extracting data from various accounts of Twin Cities history. Schmid (1937), probably the most comprehensive collection of historical statistics and facts on the growth of the Twin Cities prior to the 1940s, presented four dot graphs which illustrate the distribution of population for the two cities in the years of 1875 and 1930, respectively. Unfortunately, this data was insufficient for our analysis as no data is available beyond the city limits, or for the years between 1875 and 1930.

As direct population statistics are not available, a regional parcel data set is used in this study as a proxy. The data is maintained by Metropolitan Council annually since 2002, and the data contains about 1.07 million parcel points throughout the seven county Twin Cities metro area (Metropolitan Council, 2008). Each point feature represents a parcel that is characterized by a standard set of attribute fields, although not all attributes are completely populated. The data set includes a "year built" field which indicates the year the primary structure in a parcel was initially built. This field enables change analysis and time series investigations regarding the evolution of urban settlements in the Twin Cities. There are four "use type" attribute fields that describe the first four uses of each parcel, respectively. If parcels have more than one building, the first (primary) building in counties' list is displayed. The first field indicates the primary use type when a parcel has multiple uses. As our focus is on residential land use, residential parcels were extracted from the original data set for the analysis based on a parcel's primary use type. Another useful attribute field is "FIN_SQ_FT". This field records finished square footage of each parcel, which can be used as an approximate indicator of the intensity of residential land use.[7]

Although the "year built" field of the parcel data allows us to extract the spatial diffusion of residences in the Twin Cities in a temporal process, this data cannot be used without caution. This data is troublesome on three counts:

First, this study approximates the "snapshot" residential pattern of a particular year using currently existing residential parcels that had been built by this year. In doing so, we assumed the use type and residential area of all buildings have remained unchanged since they were initially constructed; we also ignored buildings that were torn down or replaced with new ones. Considering the intensive expansion and renewal that has taken place in the city centers, parcels in both downtowns were excluded from the analysis.

Second, residential density can be calculated as either the number of residential parcels or total finished square area of these parcels normalized by geographical area (in this case we believe the latter constitutes a better approximation of the intensity of residential land use). This calculation, though, does not differentiate an array of residential uses ranging from apartments, single-family, multi-family, to condominium. Further, only the current finished area is in the database, not the

---

[7] Unfortunately, this field is not populated for the 425,565 parcels in Hennepin County, where the city of Minneapolis is located. As a partial rescue, we acquired the Multiple Listing Service (MLS) data from Minnesota Association of Realtors. This data set contains real estate sale records during 2001-2004, which provides the finished square footage information for the parcels where a transaction occurred during this period of time. With the MLS data set we were able to fill the finished square footage information for 47,465 parcels in Hennepin, which account for over 10% of all parcels in the county.

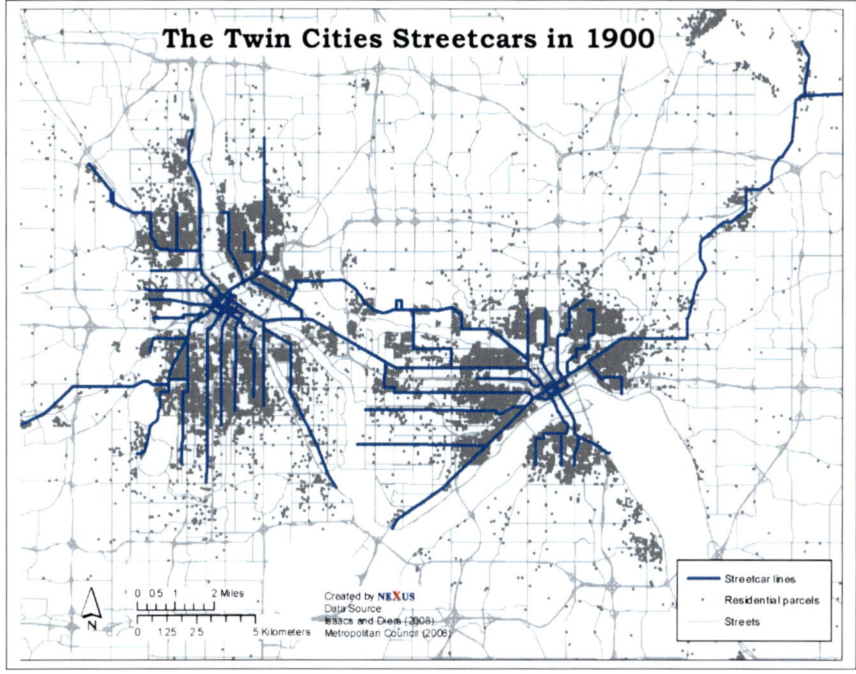

Fig. 6.1 Twin Cities streetcars in 1900

finished area at the time of construction; though the increase in finished area from the streetcar era to the present may be uniform throughout the region.

Third, a careful investigation into the parcel data set found that an abnormally large volume of parcels were coded as built in the single year of 1900. We suspect it was caused by city planners coding the dates of buildings constructed in an unknown year before 1900 as "1900" for convenience. In order to assure the accuracy of our analysis, this analysis excluded the years before 1900, and eventually covers the period of 1900-1930. The remaining data are consistent with census estimates of number of households by decade.[8]

With both streetcar and parcel data, we proceeded with replicating the temporal development of streetcar lines and residential land use in the Twin Cities. Figures 6.1, 6.2, 6.3, and 6.4 display four snapshots of the residential development coupled with the streetcar network in 1900, 1910, 1920, and 1930, respectively.

---

[8] This could be shown by regressing number of residential parcels to number of households in the seven counties with observations of 1900, 1910, 1920 and 1930. The resulting R^2 equals 0.99.

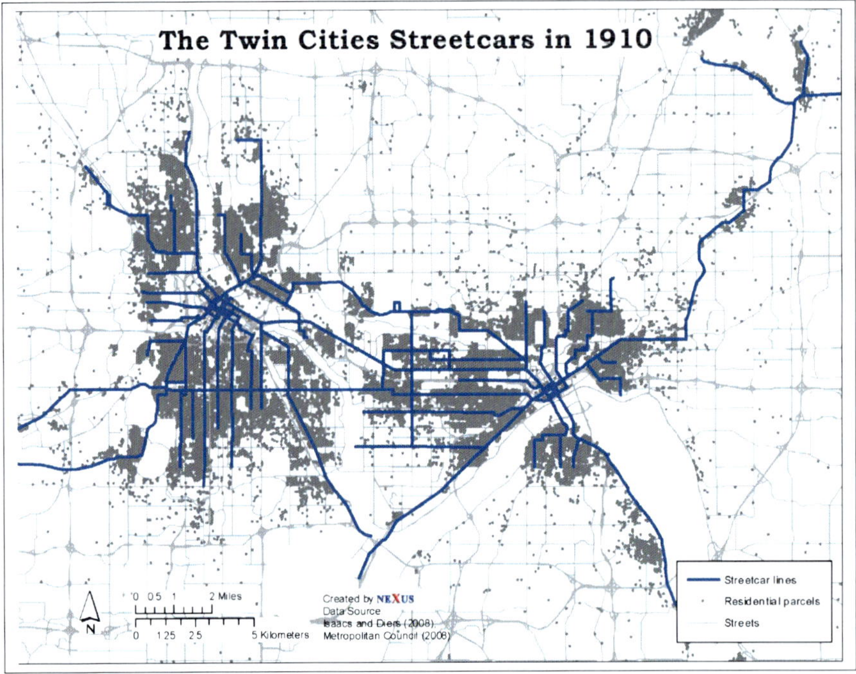

Fig. 6.2 Twin Cities streetcars in 1910

6.4 Methodology

6.4.1 Residential density vs. line density

Two models were specified to predict residential density and streetcar line density, respectively. The results will be used to test our hypothesis on the reciprocal relationship between streetcar extension and residential land use development. A cross-sectional database was constructed at a tract level to estimate the two models. The metro area was divided into 606 tracts using the 1990 census tract boundaries (the 1990 boundaries were used because the seven counties region was not completely "tracted" until the 1980s). Tracts that never had a single segment of streetcar track laid down by 1930 were excluded from our analysis; tracts that intersect either downtown were also eliminated for the previously mentioned reason. The remaining 356 tracts constitute the extent of the geographical area that was served by the streetcar network during the analysis period of 1900-1930. The cross-sectional time-series data was then constructed with observations of residential density and line density of each tract for each year between 1900 and 1930, inclusive.[9]

---

[9] The residential density of a tract in a specific year was calculated as total finished square meters of residential area constructed by the end of this year divided by the area of this tract. For those parcels

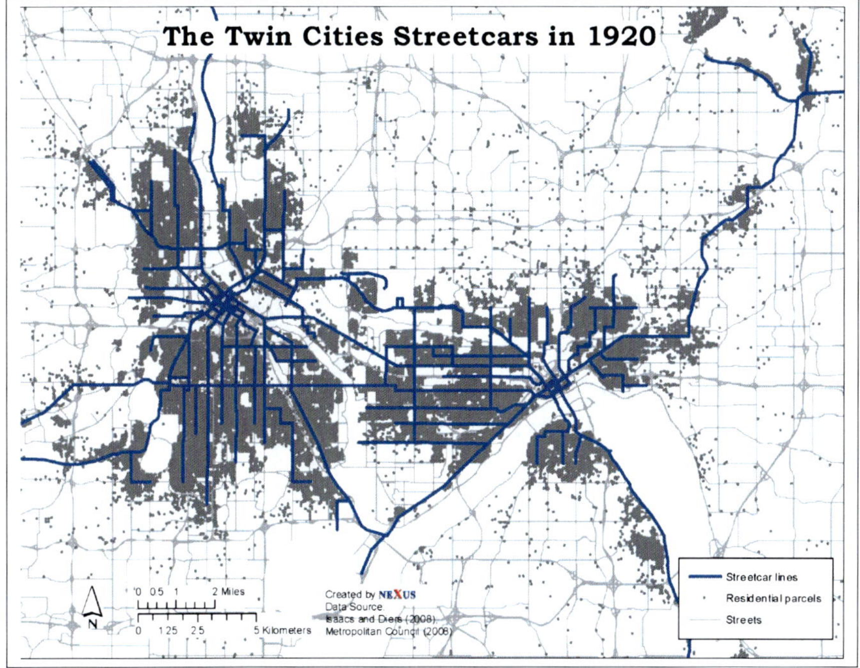

Fig. 6.3 Twin Cities streetcars in 1920

With the data set established, the residential density model is specified as follows:

$$Model\ I: R_{m,t} = \alpha_0 + \alpha_1 R_{m,t-1} + \alpha_2 \Delta L_{m,t-1} + \alpha_3 \Delta S_t + \alpha_4 D_m \qquad (6.1)$$

Where $\alpha_k, k = 0, 1, \ldots 4$ represent coefficients to be estimated. The dependent variable $R_{m,t}$ indicates the observed residential density for tract m at time t. In order for a causality test, a lag structure is included, in which $R_{m,t-1}$ indicates the residential density at time $t - 1$. The length of one time period is specified as 5 years.[10] It is expected that the residential density at time t can be largely accounted for by the lagged residential density of the same tract five years ago, so $R_{m,t-1}$ is included in the model as an explanatory variable, which also served to reduce the temporal autocorrelation in the model.

In order to test whether the addition of streetcar lines in a tract has Granger-caused the increase of residential density in the tract, a lagged change in line density

---

in Hennepin County whose finished area information were not reported, the absent information was updated with the average of the parcels within the same tract. The line density of a tract, on the other hand, was calculated as the total length of line segments opened by the end of this year normalized by tract area.

[10] A tradeoff exists between capturing more time-dependent variations with a shorter time period and reducing the computation time with a longer one. With alternatives tested, a 5-year specification accounts for the majority of the variations in an affordable computation time.

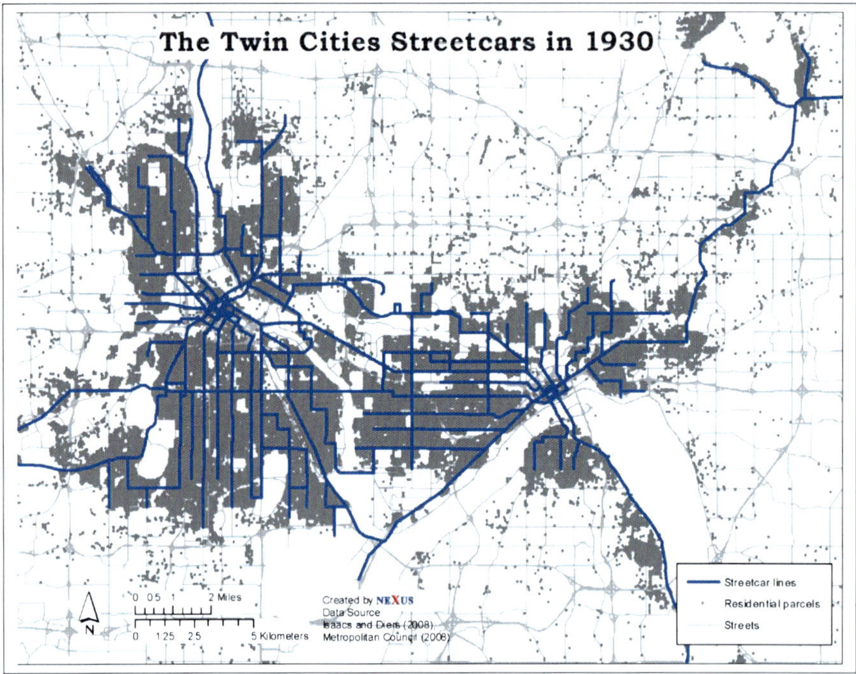

Fig. 6.4 Twin Cities streetcars in 1930

$(\Delta L_{m,t-1})$ is also included. Although two or more lags could be specified, doing so did not improve the explanatory power of the model significantly.

There are other variables that may explain the temporal or spatial variations of residential densities. Some of those variables change over time but are common to all tracts, among which this model includes total residential construction that took place in the previous periods (ΔS_t) (measured by the change of total residential area from time $t - 1$ to time t). Another set of variables is spatially differentiated but remains unchanged over time; in this model we include D_m in particular, measuring the airline distance from the centroid of each tract to the nearest downtown, assuming the locations of both city centers are fixed through time.

Similarly, another model is proposed to predict line density and to examine the causation effect in the opposite direction, namely, whether the increase of residential density had led to addition of streetcar lines. The model reads as follows:

$$Model\ II : L_{m,t} = \beta_0 + \beta_1 L_{m,t-1} + \beta_2 \Delta R_{m,t-1} + \beta_3 \Delta E_t + \beta_4 D_m \qquad (6.2)$$

Where ΔE_t indicates the total length of streetcar lines deployed in the previous time period while $\beta_k, k = 0, 1, \dots 4$ represent coefficients to be estimated.

Both models are estimated using the "xtpcse" procedure in Stata SE 10.0, which calculates panel-corrected standard error (PCSE) estimates for linear cross-sectional time-series models.

6.4.2 Proximity to line vs. residential density

Two models are proposed to test to what extent a neighborhood's proximity to the streetcar service had determined its residential density during the electric streetcar era of the Twin Cities. To conduct this test, a time-series panel data set had to be constructed at a finer geographical level (a census tract is generally too large to reflect the heterogeneity of neighborhoods within its boundary). In this case, we chose the 1990 census block boundaries, which divide the seven county metro area into 36,825 blocks. For each time period, the residential density of each block is calculated as total residential area in the block normalized by block area,[11] while proximity to the streetcar is approximated using the airline distance from the centroid of each block to the nearest streetcar line.[12] To limit the data size, observations were extracted every five years rather than every single year between 1900 and 1930, inclusive, which accounts for 7 time periods ($t = 0, 1, ..., 6$).

Model III correlates residential density of a block with its proximity to streetcars in an OLS regression. The model is initially proposed as follows:

$$Model\ IIIa : R_{n,t} = \gamma_0 + \gamma_1 D_n + \gamma_2 t + \gamma_3 P_{n,t} \tag{6.3}$$

Observations on residential density and streetcar proximity for block n at time t (indicated by $R_{n,t}$ and $P_{n,t}$, respectively) present a "snapshot" of the Twin Cities landscape in a particular time period. With data from all the seven periods combined, this model estimates the "static" relationship between residential density and proximity to streetcars in the same time period by controlling t. As it is posited that the relationship may differ in the core versus periphery of the metro area, distance to the nearest downtown is also controlled.

A correlation test, however, reveals that distance to streetcar lines and distance to the nearest downtown are highly and positively correlated (the correlation coefficient is as high as 0.92). While the close correlation between the two distance measures is itself interesting, it unfavorably caused collinearity in the model. We fixed the collinearity problem by replacing the continuous measure of $P_{n,t}$ with a set of dummy variables $\mathbf{p}_{n,t}$ which indicate proximity to the streetcar on a discrete scale. It translates into the updated model as follows:

$$Model\ IIIb : R_{n,t} = \gamma_0 + \gamma_1 D_n + \gamma_2 t + \Gamma \mathbf{p}_{n,t} \tag{6.4}$$

Where $\mathbf{p}_{n,t}$ represents a set of 9 dummy variables and Γ a vector of coefficients before the dummies. The dummy variables determine if a block is located within one of the following 9 distance ranges: 0-400, 400-800, 800-1200, 1200-1600, 1600-2400, 2400-3200, 3200-4800, 4800-6400, 6400-8000 (with all distances measured

---

[11] A block is dropped from the data if none of its parcels have residential area information.

[12] It should be noted that in reality people generally walked along streets and alleys to access streetcars. So network distance may be a more accurate measure of the actual proximity (we expect the numbers to be generally proportional though) Airline distance, however, is adopted because it is much easier to calculate while still reflecting the relative distance away from streetcar lines. This distance is calculated using the "near" analysis in ArcGIS.

in meters). This essentially allocate all the blocks into 10 ring buffers according to their relative proximity to streetcar lines. Each dummy variable is denoted by the upper-bound value of its corresponding distance range. For instance, $p_{n,t}^{6400}$ equals 1 if block n is located 4800-6400 meters within the streetcar service at time t, and equals 0 otherwise. If a block is more than 8000 meters away from any streetcar line, all the dummies will equal 0.

The next model is specified as follows to examine whether the change in proximity to the streetcar service precedes the change in residential density:

$$Model\ IV : R_{n,t} = \eta_0 + \eta_1 R_{n,t-1} + \eta_2 \Delta S_t + H\Delta \mathbf{p}_{n,t-1} \tag{6.5}$$

To capture the potential causation effect, the model includes residential density of a block $(R_{n,t})$ as the dependent variable while the lagged changes in proximity measures as explanatory variables on the right-hand side $(\Delta \mathbf{p}_{n,t-1})$. Note that the change of a dummy variable equals 1 when distance to line is shortened to the corresponding distance range, and equals 0 when distance to the line remains unchanged or changes within the range. Since no streetcar lines had been dropped by 1931, the temporal change in proximity measures will never be negative in our analysis period. This model also includes the change of total residential area (ΔS_t) as an explanatory variable.

As in the tract analysis, blocks that intersect either downtown are eliminated. In order for a balanced time-series data, blocks whose residential density observations are incomplete (missing data in any of the seven periods) are also excluded. Finally a data set of 30,310 records (4,330 blocks by 7 time periods) is obtained, which is still too large for a panel-corrected standard error analysis in Stata. This issue was addressed by randomly selecting 20% (866) of all eligible blocks for the analysis.

6.5 Results

The results from *Model I* and *Model II* are presented in Tables 6.1 and 6.2, respectively. As shown in Table 6.1, residential density $(R_{m,t})$ of tract m at time t is highly and positively correlated with five-period (five years) lagged streetcar line density $(\Delta L_{m,t-1}.L5)$ of this tract, suggesting the extension of streetcar lines (increase in line density) has caused a significantly increased volume of land devoted to residences five years hence. This finding corroborates our hypothesis that the deployment of streetcar lines, especially those extending into open vacant suburban land during the electric streetcar era, had brought new residents and new housing to the area they reached. In contrast, the lagged residential density $(\Delta R_{m,t-1}.L5)$ is not a significant predictor of streetcar line density. Results from the two models suggest that the expansion of the streetcar system had to a significant extent led residential construction, and profoundly shaped the Twin Cities landscape. The results also support our hypothesis that unlike in those transportation systems whose infrastructure investment was made mainly in response to increased demand, the rapid expansion of the Twin Cities streetcar system outside of downtown was largely attributed to other factors such as superiority of technology, TCRT's monopoly ownership and close connections with real estate business, and travelers' reliance on streetcars during that era.

Table 6.1 Model I: Predicting tracts' residential density using lagged changes in line density

Number of observations 3234
Number of groups   154
Observations per group   21
R^2 0.95
Wald χ^2 2249
Dependent variable $R_{m,t}$

Explanatory Variables	Coeff.	Std.Err.	P-Val.
Lagged residential density ($R_{m,t-1}$) (L5)	1.06	0.03	0.00
Lagged change in line density ($\Delta L_{m,t-1}$) (L5)	2.93	0.83	0.00
Change in total residential area (ΔS_t)	5.68E-10	3.78E-10	0.13
Distance to nearest downtown (D_m)	1.38E-07	7.99E-08	0.08
Constant	5.91E-03	1.32E-03	0.00

Table 6.2 Model II: Predicting tracts' line density using lagged changes in residential density

Number of observations 3234
Number of groups   154
Observations per group   21
R^2 0.94
Wald χ^2 11736
Dependent variable $L_{m,t}$

Explanatory Variables	Coeff.	Std.Err.	P-Val.
Lagged line density ($L_{m,t-1}$) (L5)	0.96	0.01	0.00
Lagged change in residential density ($\Delta R_{m,t-1}$) (L5)	-2.94E-04	6.55E-04	0.65
Change in total line length (ΔE_t)	4.49E-09	2.71E-10	0.00
Distance to nearest downtown (D_m)	-1.02E-08	2.66E-09	0.00
Constant	1.03E-04	3.45E-05	0.00

Table 6.3 displays the results from *Model IIIb*, which tests the "static" relationship between residential density and proximity to streetcar lines in the same time period. Not surprisingly, residential density declines with distance to downtowns,

implying much more intensive land development in the city centers than in the suburbs. The positive coefficients before time period (t) suggest residential density increases over time. With these two variables controlled, coefficients of proximity measures (dummies) are all statistically significant, and decline as the distance value increases, in other words, the nearer to streetcar lines, the more residential area developed. The coefficient before $p_{n,t}^{400}$, for instance, implies that per 10,000 square meters of land, 461 more square meters of residential area were constructed 400 meters within the streetcar lines than beyond 400 meters. Dummies within the distance of 1,200 meters are all positively correlated with residential density, while those beyond 1,200 meters are negatively correlated, meaning if we divided residences into two groups within and beyond a particular distance to streetcar lines, 1,200 meters is the threshold distance that would see the biggest drop in residential density away from a streetcar line. This agrees with our hypothesis that residential density will significantly drop beyond a workable distance to access the streetcar, due to the lack of complementary transportation that would feed the streetcar service. This finding explains the historical observation of "finger-shaped" residential distributions in the Twin Cities. It is also consistent with the modern notion about people's willingness to walk to transit (usually said to be 400 or 800 meters). O'Sullivan and Morrall (2007), for example, found for the city of Calgary, Canada the average walking distance to suburban stations is 649 meters with a 75th-percentile distance of 840 meters, and that to CBD stations is 326 meters with the 75th-percentile distance of 419 m.

Table 6.3 Model IIIb: Correlating blocks' residential density with streetcar proximity measures

Number of observations 252448
R^2 adjusted   0.25
Root MSE   0.03
Dependent variable $R_{n,t}$

Explanatory Variables	Coeff.	Std.Err.	P-Val.
Distance to nearest downtown (D_n)	-1.41E-07	7.90E-09	0.00
Time period (t)	1.68E-03	3.23E-05	0.00
Proximity to line dummies			
Dummy 0-400m ($p_{n,t}^{400}$)	4.61E-02	2.72E-04	0.00
Dummy 400-800m ($p_{n,t}^{800}$)	1.88E-02	3.21E-04	0.00
Dummy 800-1200m ($p_{n,t}^{1200}$)	4.62E-03	3.60E-04	0.00
Dummy 1200-1600m ($p_{n,t}^{1600}$)	-6.57E-04	3.89E-04	0.09
Dummy 1600-2400m ($p_{n,t}^{2400}$)	-3.21E-03	3.28E-04	0.00
Dummy 2400-3200m ($p_{n,t}^{3200}$)	-4.09E-03	3.50E-04	0.00
Dummy 3200-4800m ($p_{n,t}^{4800}$)	-4.22E-03	2.84E-04	0.00
Dummy 4800-6400m ($p_{n,t}^{6400}$)	-3.59E-03	2.98E-04	0.00
Dummy 6400-8000m ($p_{n,t}^{8000}$)	-3.72E-03	3.07E-04	0.00
Constant	1.65E-03	2.53E-04	0.00

The results from *Model IV* are reported in Table 6.4. It should be noted that one-period lag (L1) in this model represents 5 years of lag (which is consistent with *Model I* and *Model II*), as the block data was extracted every 5 years. As can be seen, the presented model includes only dummies with positive coefficients in *Model IIIb*. This model presents the highest explanatory power among variations that include different sets of dummy variables. As Table 6.4 shows, residential density is positively correlated with the lagged changes in streetcar proximity measures, suggesting construction of a new line will Granger-cause an increase in residential density. The causation effect, however, is not significant beyond 800 meters (as the lag of $\Delta p_{n,t-1}^{1200}$ is not statistically significant in predicting residential density). This further corroborates our observation that proximity to the streetcar has a significant effect on residential development only within a walkable distance. Although both exhibit strong causation effects on residential density, it is a little surprising that gaining access to streetcar lines within 400 meters ($\Delta p_{n,t-1}^{400}$) is not as significant a predictor as gaining access within 400-800 meters ($\Delta p_{n,t-1}^{800}$). This might be explained considering that our analysis includes observations only during 1900-1930 due to the aforementioned data issue, while land development might have taken place substantially within 400 meters of streetcar lines during the first decade of the electric streetcar era (1889-1899). The first 400 meters might also be disproportionately commercial, as streetcars tended to run along commercial streets (and the streets along which streetcars ran tended to become commercial if they were not already).

Table 6.4 Model IV: Predicting blocks' residential density using lagged changes in line proximity

Number of observations 4330
Number of groups 866
Observations per group 5
R^2 0.95
Wald χ^2 2153
Dependent variable $R_{n,t}$

Explanatory Variables	Coeff.	Std.Err.	P-Val.
Lagged residential density ($R_{n,t-1}$)(L1)	1.04	0.04	0.00
Lagged change in line proximity dummies			
Dummy 0-400m ($\Delta p_{n,t-1}^{400}$) (L1)	5.18E-03	3.28E-03	0.11
Dummy 400-800m ($\Delta p_{n,t-1}^{800}$) (L1)	5.42E-03	2.78E-03	0.03
Dummy 800-1200m ($\Delta p_{n,t-1}^{1200}$) (L1)	2.40E-03	2.11E-03	0.25
Change in total residential area (ΔS_{t-1})	4.38E-08	1.24E-08	0.00
Constant	2.01E-03	9.82E-04	0.04

6.6 Findings and concluding remarks

In contrast with Chapters 4 and 5 that treat land use as exogenous, this research analyzes causation in the coupled development of population and streetcars in the Twin Cities metropolitan area. Historic residence and network data were constructed between 1900 and 1930, and linear cross-sectional time-series models were estimated at both a tract and block level using this data. Granger causality tests disclose that while the extension of electric streetcar lines preceded intensive residential development where streetcars reached, population growth did not predict line additions. This leads to the conclusion that unlike those transportation systems which were expanded in response to increased demand, the rapid expansion of the streetcar system during the electric era has been driven by other forces and to a large extent led land development in the Twin Cities. The main forces that have driven this process include superior technology, monopoly, real estate development, and people's reliance on the streetcar for mobility.

Lacking alternative public transportation, proximity to streetcar lines became a crucial factor that determined where people live in that era. In analyzing the relationship between residential density and streetcar proximity, we observed that residential density declines with the distance from streetcar lines, and significantly drops beyond a walkable distance; we also observed that attaining streetcar service within 800 meters (about a half mile) predicted the increase in residential density to a significant extent.

The implications of this research are several. First, the case of Twin Cities streetcar system contributes to a deeper understanding of the supply-demand interaction in transportation. While the reciprocal relationship between population growth and transportation investment is pervasive, there are occasions when one significantly leads the other while the inverse process does not hold. The reasons could be technological, economic, managerial, social, or political. In fact, each urban transportation system has a unique course of development that is shaped by multifaceted forces. Second, the recognition of "induced demand" and "induced supply" effects are important for city and transportation planners to better understand and regulate urban transportation systems from a long-term perspective. Granger causality tests provide an analytic tool that can assist in examination of the presence and strength of both effects. Finally, this study presented empirical models that account for the temporal and spatial variations of population and transportation network in an integrated process, which, complementary to the preceding studies that treat land use as exogenous, calls upon the consideration of transportation and land use interaction in modeling the growth of transportation networks.

Chapter 7
First Mover Advantages

7.1 Introduction

The use of a fairly standard vehicle width of a little under 2 meters originates from the design of prehistoric carts and sleds as evidenced by rutting in ancient roads which aided in steering. Despite dramatic advances in vehicular and infrastructural technologies, the standard has changed little over the millennia. The gauges of railroad track, for instance, are now standardized at 4 feet 8 and half inches (1435 mm) across Europe and North America, the same as the first steam railway, and a mere half-inch wider than the typical pre-steam tracks in the mining districts near Newcastle, consistent in size with the wheel gauge used in Roman Britain. This standard gauge lasted since it was first used on the Stockton and Darlington railway in 1825, and were adopted by most subsequent lines (Puffert, 2002), despite some railways trying alternatives (e.g. the Great Western Railway was originally built at 5 feet 6 inches, or 1676 mm). Alternative gauges would have accommodated wider, taller, and faster trains more easily, but the standard gauge that was adopted first acquired advantages as other railways sought compatibility with the standard to obtain access to the uses of earlier lines, and helped lock-in that standard. Tracks of alternative gauges could not be deployed economically because of the lock-in, including requirements to rebuild expensive sunk infrastructure like bridges and tunnels to accommodate the wider gauge.

The standardization of railway gauges provides an example of first mover advantages (FMA)[1] conferred by standard lock-in. Similarly, a technology may be adopted before others and acquire advantages as this technology is diffused and locked in. A theory of technology diffusion suggests that technologies are deployed

---

[1] The notion of first mover advantage (FMA) probably stemmed from ancient chess or other competitive board games, where the player who makes the first move has an inherent advantage to win the game (Streeter, 1946). Since the introduction of modern game theory in the 1940s (von Neumann and Morgenstern, 1944), first mover advantage has developed into a game theoretic notion that, in a sequential round of strategic moves, a player may earn greater pay-off being the first to act rather than following others. Examples include the two-stage Stackelberg Game and Cournot Game(Gibbons, 1992).

in a pattern resembling an S-shaped curve (Kondratieff, 1987). There is a long period of birthing, as the technology is researched and developed, there is a growth phase as the technology is deployed, and a slower mature phase as the technology has occupied available market niches. Nakicenovic (1998), by plotting a large number of curves for major transportation modes, showed that S-curves fit the temporal realization of transportation infrastructure networks very well.

Besides standard and technology lock-ins, another important source of first mover advantages arises from spatial heterogeneity. Rather than a system that can be just deployed universally, transportation infrastructure must be deployed in some place first. First deployed transportation infrastructure can occupy locations where it is most demanded or most cost-effective to build. Furthermore, transportation infrastructure usually embeds high set-up and sunk costs, which make it more difficult for the incumbent to move, as they are physically bound to the location in which they have sunk costs, but they also help an incumbent if it is well located by establishing spatial monopoly and chasing off rivals. Thus infrastructure deployed in a particular place may gain an advantage because it adopted a standard / technology first in time, or because it acquired the best location, or both. Just as embeddedness may help a first mover attain or retain its position, it also keeps systems from evolving beyond those positions (Hommels, 2005).

In the last decade, network scientists have revealed that in many complex systems, such as the world-wide web and metabolic networks, nodes "entering the system at the early times have always the largest connectivities and strengths" as those with higher degree of connections have stronger ability to grab links subsequently added to the network (Newman, 2003; Barrat et al., 2004). Although surface transportation networks differ from these node-centric networks due to spatial constraints, this concept of preferential attachment sheds new light on how advantages of first movers can be established and reinforced in an evolutionary process of network growth.

This chapter aims to examine the existence or absence of first mover advantages in the transportation sector. Four empirical examples illustrate qualitatively and quantitatively different aspects of first mover advantages in the development of modern transportation systems. These examples include rail in London, the global aviation and maritime systems, and roads in the Twin Cities of Minneapolis and Saint Paul, discussed in turn.

7.2 Rail in London

The steam railway reached Greater London in 1836. Its extent is defined as the number of London area stations. It reached 50 percent of its ultimate extent (excluding stations that were later closed) in 1868 and 90 percent in 1912. The first Underground railway was opened in London in 1863, connecting stations of the various surface railways. Half of the ultimate number of stations were open by 1912, and 90 percent by 1948. These data are illustrated in Figure 7.1.

Figure 7.1 shows two graphs, one each for the National Rail (surface) lines and for the London Underground. Each graph contains three lines. The first is the cumulative share of the number of stations opened by year. For the London Underground, 0 percent of stations were open in 1862, and 100 percent were open by 1999. The second is the cumulative share of number of current boardings and alightings at stations by the year they opened. So the number for a given year represent the current boardings and alightings for stations that were open by that year. The third indicates the share of the number of connections (in the case of National Rail) or the number of lines using that station (in the case of the London Underground) by year.

What is apparent by observing the graphs is that the share of cumulative ridership is greater than the share of cumulative connections, which is greater than the share of cumulative stations. In other words, the early stations have more connections than later stations, and more still riders than later stations.

This would support the notion of a first mover advantage. The early stations were generally well placed in the areas that at the time generated more traffic. While land use patterns and demand have shifted in London (Levinson, 2008*b*), the underlying pattern was of early stations serving the then dense core. The core sustains as a dense employment center in London. The early stations, those in the core, are also more likely to have multiple connections, but the additional connections do not of themselves explain the additional ridership, rather we need to look outside the network at the land use, and the mutual reinforcement between the land use and the network, as an explanation (Levinson, 2008*a*).

Population data was obtained for the 33 current Administrative Districts (also called Boroughs, including the City of London and City of Westminster) of London. Density of population (and employment) were computed by dividing by the current area. This paper defines the surface rail system as all currently existing London area heavy rail stations and lines that are not part of the 2006 Underground system, and the Underground stations are those that are part of the 2006 Underground system. Transport network data on the London Underground identifies each station on each line as a node, with X and Y coordinates, a date opened for a particular line. Very few stations were actually closed (as opposed to relocated), and these did not result in notable service reductions, but instead were because stations were positioned too close together initially. Underground stations that were opened and later closed are not considered as part of the analysis. Dates were obtained from Rose (1983) and Borley (1982). Small relocations of stations are ignored, as is the Circle Line, which shares platforms with the District, Metropolitan, or Hammersmith and City lines. Once a new line serves the station, that new line at the station is given a new node for the purposes of station density, but each station (which may serve multiple lines) remains a single observation. The density of Underground stations was computed by dividing the number of Underground nodes (each station per line is treated as a distinct node, so a station serving three lines is counted as three nodes) at a given time by the current area. A similar procedure is used for surface rail stations.

Examining the data more rigorously in Table 7.1 suggests the source of the first mover advantage is the spatio-temporal location on the network and connectivity. Total station boardings and alightings on both the surface rail and Underground

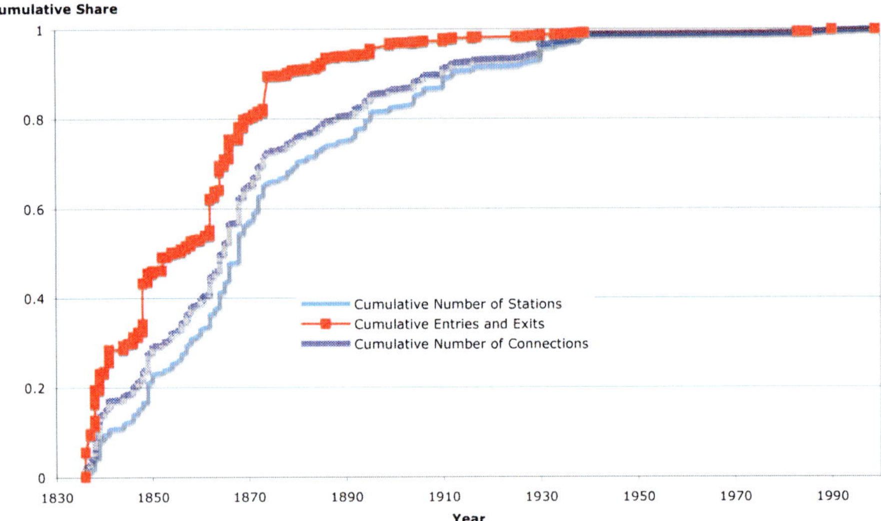

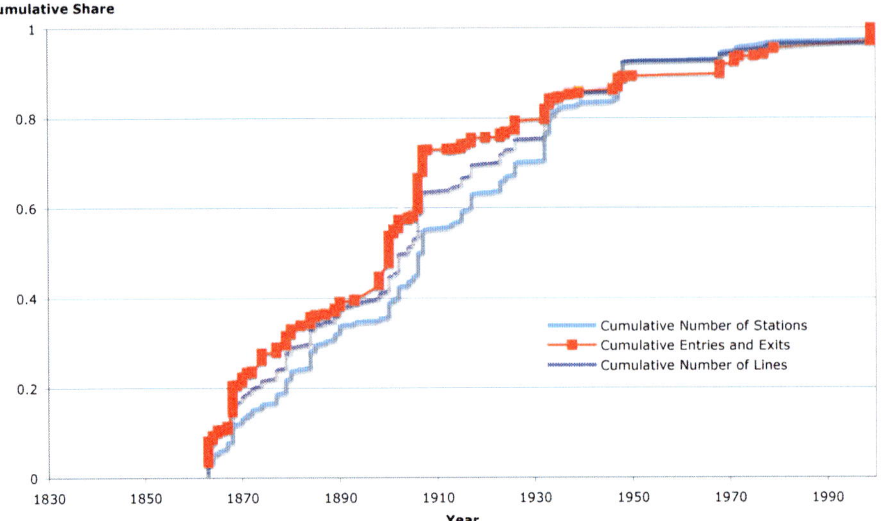

Fig. 7.1 London rail cumulative stations and ridership. The first stations serve more riders than later stations. The first stations also have more connections than later stations, but not so much as to explain the additional ridership.

networks are positively related to number of connections and negatively related to travel time to Bank station (in minutes) (roughly at the center of the City of London) and statistically unrelated to year after controlling for those two variables. Other variables controlled for are population and employment density (in thousands per km^2) (which were insignificant, though notably highly correlated with station density), station density (in stations per km^2) (Underground station density was insignificant, surface rail station density was negative in both models, suggesting surface rail stations compete for customers while Underground stations complement, perhaps through higher densities), and location north of the Thames River (surface rail stations are somewhat less successful north of the Thames) and in the urban core (the London Boroughs of City of London, Westminster, Camden, Islington, Tower Hamlets, Kensington and Chelsea, and Southwark.

The core is defined as having a high degree of employment, areas where the ratio of persons working in the area to working-age residents exceeds one, (values in parentheses) (City of London (55.74), Westminster (3.65), Camden (1.84), Islington (1.38), Tower Hamlets (1.16), Kensington and Chelsea (1.08), and Southwark (1.02)). These areas are seven of the eight boroughs of London that have a job to working age population ratio greater than one. (The other area is Hillingdon (1.16), at the edge, the borough in which Heathrow Airport is located, and so is otherwise dissimilar from the core and is considered part of the periphery here) (Center for Economic and Social Inclusion, 2006).

The entrepreneurs developing the rail system placed early stations well to take advantage of existing and prospective demand; the value of that placement remains today, a century after most stations opened.

Table 7.1 London rail boardings and alightings regression model

Independent Variables	Surface Rail Stations			Underground Rail Stations		
	Coeff.	T-Stat	P	Coeff.	T-Stat	P
Year	-3829	-0.4	0.69	25881	1.55	0.12
No. of connections	1850620	10.45	0.00 ***	9136712	10.27	0.00 ***
Population density	160836	0.84	0.40	446271	1.46	0.15
Employment density	218436	1.58	0.12	291440	1.42	0.16
Underground station density	2391224	1.15	0.25	2250592	0.83	0.41
Surface rail station density	-7369758	-4.17	0.00 ***	-4561511	-2.45	0.02 **
North of Thames	-1109628	-1.79	0.08 *	-1197065	-0.67	0.50
Core	-1556244	-1.26	0.21	112884	0.06	0.95
Time to Bank station	-116756	-3.8	0.00 ***	-211722	-3.37	0.00 ***
constant	9077731	0.51	0.61	-49200000	1.49	0.14
Adjusted R-squared	0.5339			0.4759		
N	308			257		

7.3 Aviation

The world aviation system allows us to test first mover advantage under particular conditions. There are a number of measures of airport size, one is number of passengers, as shown in Table 7.2.

If there were a first mover advantage, we would expect to see the oldest airports as the largest. The table indicates otherwise, among the world's largest airports, there is no particular advantage to being a city with an earlier airport. Amsterdam-Schiphol which opened in 1916 is ranked 12, while Dallas-Fort Worth (DFW), opening in 1973, is ranked 6. One could argue that DFW is best considered a successor to an older airport, Love Field in Dallas opened in 1917 and opened for civilians in 1927 while Meacham Field in Fort Worth opened in 1925. Those airports were different institutions located on different sites though. Still, Dallas (along with Chicago) was established as a hub of American Airlines as early as 1930 (and predecessor companies were already using that airport).

In contrast with first mover advantages when comparing large airports, we can see significant persistence of hubs. Airlines that establish hub airports (a) don't move them very often, and (b) crowd out other airlines seeking to establish hubs. For instance, Northwest Airlines was established in Minneapolis in 1926 and remains the dominant airline in that market eighty years later (under the name of its successor Delta). American Airlines in Dallas and Chicago tell similar tales. This persistence is not determinative, airlines with hubs do disappear (Eastern Airlines is a notable example), and do lose their hub advantage when faced with strong competitors. The case of US Airways versus discount carrier Southwest Airlines in Philadelphia is a telling example, when the former CEO David Siegel of the then entrenched US Airways said "They are coming to kill us", foretelling the loss of market share in Philadelphia (Business Travel News Online, 2004), after Southwest had taken the Baltimore and west coast markets from them. US Airways subsequently merged with America West, which took the US Airways name, but headquartered itself in Phoenix.

Within a metropolitan area, it is possible there is a first mover advantage, as the first airport might capture the dominant or monopoly share of locally generated traffic. However, that is difficult to test as so many cities have only one airport, and cities that once had more may have relocated their traffic to a more suitable location. Alternatively, there might be a second mover advantage as a newer airport may be better adapted to the changed environment than earlier facilities, which might be on sites that are too small or face high costs of rebuilding in place while remaining operational.

The surviving large airlines (network carriers) in the US aviation system can trace their heritage to before the jet age, when air travel was uncommon and largely subsidized by airmail contracts from the post office. Each airline has a distinct history, and the industry is rife with consolidation. American airlines can trace its heritage to some 72 precursor companies. To illustrate with a simpler example, Northwest merged with Republic airlines in 1986, Republic itself was the product of a 1979 merger between North Central Airlines (based in Minneapolis though founded in

Table 7.2 Airport passengers in 2006 (Top 30) by Opening Year

Rank	Airport	Passengers	Open Year
1	Hartsfield-Jackson Atlanta	84,846,639	1925
2	O'Hare	76,248,911	1942
3	London Heathrow	67,530,223	1946
4	Tokyo (Haneda)	65,225,795	1931
5	Los Angeles	61,048,552	1929
6	Dallas-Fort Worth	60,079,107	1973
7	Charles de Gaulle	56,808,967	1972
8	Frankfurt	52,810,683	1936
9	Beijing Capital	48,501,102	1958
10	Denver	47,324,844	1989
11	McCarran	46,194,882	1942
12	Amsterdam Schiphol	46,088,221	1916
13	Madrid Barajas	45,500,469	1928
14	Hong Kong	44,020,000	1998
15	Suvarnabhumi	42,799,532	2006
16	George Bush Intercontinental	42,628,663	1969
17	John F. Kennedy	42,604,975	1948
18	Phoenix Sky Harbor	41,439,819	1935
19	Detroit - Wayne County	36,356,446	1930
20	Minneapolis-Saint Paul	35,633,020	1921
21	Newark Liberty	35,494,863	1928
22	Singapore Changi	35,033,083	1955
23	Orlando	34,818,264	1974
24	London Gatwick	34,172,489	1936
25	San Francisco	33,527,236	1927
26	Miami	32,533,974	1928
27	Narita	31,824,411	1978
28	Philadelphia	31,766,537	1925
29	Toronto Pearson	30,972,566	1939
30	Soekarno-Hatta	30,863,806	1984

Source: Airports Council International (2006)

Wisconsin in 1939 and not moving to Minneapolis until 1952) and Southern Airways (founded 1949 in Augusta, Georgia), and a 1980 acquisition of Hughes Airwest (the product of a 1968 merger between Pacific Airlines (founded 1941 in California as Southwest Airways), Bonanza Air Lines (founded 1945 in Las Vegas) and West Coast Airlines (founded 1946 and based in Seattle). The hub of Northwest at Minneapolis was the first airport served by the original Northwest, while the original hubs or bases of predecessor airlines are no longer the dominant hubs of the current Delta.

Further we could investigate the other way, what were the first hub airports of the six airlines. The columns in Table 7.3 showing the first airmail and first passenger routes give insight into the first markets airlines occupied. These markets were allocated by government (either the post office giving airmail contracts or the

Table 7.3 US network airline hub cities

Airline	Year	Hub Cities	First mail service	First passenger service
American Airlines	1930	**Dallas**, Miami, San Juan, **Chicago**, St. Louis	St. Louis, Chicago	Dallas, Chicago, Boston
United Airlines	1926	Chicago, Denver, Washington (IAD), San Francisco, Los Angeles	Boise, Pasco	Chicago, Kansas City, Dallas
Delta Airlines	1924	Atlanta, Cincinnati, *Salt Lake City*, New York (JFK)	Fort Worth, Atlanta, Charleston	Dallas, Jackson
Continental	1934	Houston, *Newark*, Cleveland		El Paso, Las Vegas, Albuquerque, Santa Fe, Pueblo
Northwest Airlines	1926	**Minneapolis**, Detroit, Memphis, Tokyo, Amsterdam	Minneapolis, Chicago	Minneapolis, Chicago
US Airways	1939	*Charlotte*, Philadelphia, *Phoenix*, Las Vegas, **Pittsburgh**	Pittsburgh	Pittsburgh

Source: Airline websites

Note: **bold** indicates original airport served by airline. Other hubs were often served by acquired companies, *italics* indicates original airport served by an acquired company.

Civil Aeronautics Board allowing airlines to serve passenger markets). Those original markets are in five of six cases (Continental excepted) still dominated by the successor airline, and Continental moved first to Denver, which it eventually pulled back from and then east to Houston (especially after acquisition by Texas Air Corporation), a hub which it dominated.

Airlines, unlike airports, are largely comprised of mobile capital. Despite the mobility of the main capital asset, airplanes, there is a persistence of airlines at hubs (five of the six US network carriers retain at least one hub from their earliest days), and one considers both passenger and mail service, all six airlines are still serving at least one of those markets, shows there is a persistence of airlines at hubs. The airline industry, because of its regulated nature through the 1970s, was a product of merger and consolidation as much as internal growth. The lock-in advantages, in addition to those noted by other authors above, include ownership or control of scarce gates at competitive airports, frequent flyer programs which tie local residents to the local airline, and hubbing economies (a type of network effect) allowing hubs to have frequent non-stop service to many cities.

7.4 Container ports

Until containerization, longshoremen climbed onto cargo ships and using brute force, with the aid of nets and grappling hooks, moved small packages on and off ships. The process of loading and unloading might keep a ship in port for weeks. This break-bulk shipping was a major bottleneck in world commerce.

Malcolm McLean, a trucker from North Carolina, conceived of loading trucks directly onto ships, without packing and unpacking, using ships as transoceanic ferries. If the wheels were removed and the sides reinforced, trailers could be stacked. In April 1956 McLean's first container ship sailed from New York to Houston.

Containerization was essentially complete in 1971, when all containerizable cargo on the trans-Atlantic route was containerized (Rosenstein, 2000) . Yet the revolution continued both as the quantity of shipped freight and the size of the ships (and the ports required to load them) grew.

The scaling made many older, smaller ports obsolete and created a generation of new superports that acted as hubs in a packet-based freight transportation system. Table 7.4 shows Container Port Size in 1969, near the beginning of containerization. One notes for instance that Oakland had already beaten its competitor across the bay in San Francisco to containerization. Table 7.5 shows container port size in 2005, and a different picture emerges, only four of the top ten ports in 1969 (denoted in bold in both tables) remained in the top twenty, only two in the top ten. Oakland, the number two largest container port in 1969 fell out of the top twenty as Los Angeles rose to take US west coast market share. Australia's ports of Sydney and Melbourne also fell off the list, Port of Yokohama, Japan was displaced by the slightly larger neighboring Port of Tokyo on the list, Bremen was replaced by Hamburg, and Felixstowe (serving southeast England) also fell off the list.

The entrants on the list are all from East or Southeast Asia, and Dubai which has emerged to fulfill a transshipment role for the Middle East.

What does this say about first mover advantages? Ports are immobile capital, while a port is certainly an important aspect of a city's growth, it cannot alone determine that growth. As city-regions grow, and some specialize in producing or distributing tradable goods suitable for containerization, those ports will similarly grow. A port that grows early may retain some disproportionate advantage for a time while equilibrium is established, and by doing so, may provide advantages for other complementary aspects of manufacturing and trade helping to enmesh its position. The evidence however suggests that first mover advantages are quite weak in this sector.

7.5 Roads in the Twin Cities

The Minnesota Department of Transportation (and predecessor organizations) have been building and maintaining roads in the Twin Cities region since 1921. For the purpose of this study, we have assembled a database of road projects by section, and analyzed the relation between the year they were built, and their current utilization (measured here as annual average daily traffic, or AADT). The results, shown in Table 7.6, indicate that the later the year, the greater the AADT, implying the more recently constructed links carry more traffic. This observation holds for state

Table 7.4 Container port size in 1969

Rank	Port	Container Cargo (Metric tons)
1	**New York/New Jersey**	4,000,800
2	Oakland	3,001,000
3	**Rotterdam**	2,043,131
4	Sydney	1,589,000
5	**Los Angeles**	1,316,000
6	**Antwerp**	1,300,000
7	Yokohama	1,262,000
8	Melbourne	1,134,200
9	Felixstowe	925,000
10	Bremen/Bremerhaven	822,100

Source: Levinson (2006)

Table 7.5 Container port size in 2005

Rank	Port	TEUs(000s)
1	Singapore	23,200
2	Hong Kong	22,430
3	Shanghai	18,090
4	Shenzhen	16,200
5	Busan	11,840
6	Kaohsiung	9,471
7	**Rotterdam**	9,300
8	Hamburg	8,086
9	Dubai	7,619
10	**Los Angeles**	7,485
11	Long Beach	6,710
12	**Antwerp**	6;325
13	Qingdao	6,307
14	Port Kelang	5,544
15	Ningbo	5,208
16	Tianjin	4,801
17	**New York/New Jersey**	4,793
18	Tanjung Pelepas	4,169
19	Laem Chabang	3,766
20	Tokyo	3,594

Source: Port of Hamburg (2005)

routes and US highways, which were both largely planned in an *ad hoc* way, but not for interstate highways, which were more centrally planned, and for which year is insignificant.

Table 7.6 Traffic on highways in Minnesota

	State Highways			US Highways			Interstate Highways		
	Coeff.	*T-Stat*	*P*	*Coeff.*	*T-Stat*	*P*	*Coeff.*	*T-Stat*	*P*
const.	-1097804	-3.46	0.0009	-1267179	-5.16	0.0000	1688215	0.83	0.41
year	581	3.55	0.0007	674	5.32	0.0000	-810	-0.78	0.44
Adj. R^2		0.14			0.46			0	
N		74			33			29	

A number of factors below have been found to explain the fact that more recently constructed highways carry more traffic, which may or may not suggest disadvantages of first movers:

First, as technologies of road construction and pavement continuously improve, later-built links may be able to adopt a new technology (e.g., divided highways functionally replace undivided highways, grade separation functionally replaces traffic signals), which increases operating speed, and attracts more traffic. On the other hand, some strategic links, such as bridges or tunnels, may be technically difficult to construct, thus appearing late despite their importance in location. Though our analysis did not consider the technological factors.

Second, highways may have been constructed to replace local streets and arteries, which unfortunately we were not able to include in our data set. A heavily traveled highway may be constructed recently, but its predecessor road which it replaced may have been there much earlier.

Last but not least, the Twin Cities metropolitan area has experienced rapid but spatially uneven demographic development over decades. Figure 7.2 displays population by county in the Twin Cities seven-county metro area from 1920 to 2000. While a significantly larger share in total population is seen in suburban counties like Anoka, Dakota, and Washington, a shrinking share is observed in the more urban Hennepin and Ramsey. Uneven development leads to the shift of spatial agglomeration. Consequently, important locations decades ago may become less important and even obsolete over time, and so do the links that were built to connect these locations.

The case of Twin Cities roads suggests that first mover advantages be analyzed in an evolutionary context. First mover advantages of transportation links (or routes) cannot be divorced from the temporal development of networks and places. Empirical analyses, however, are usually limited by the availability of extensive historical data.

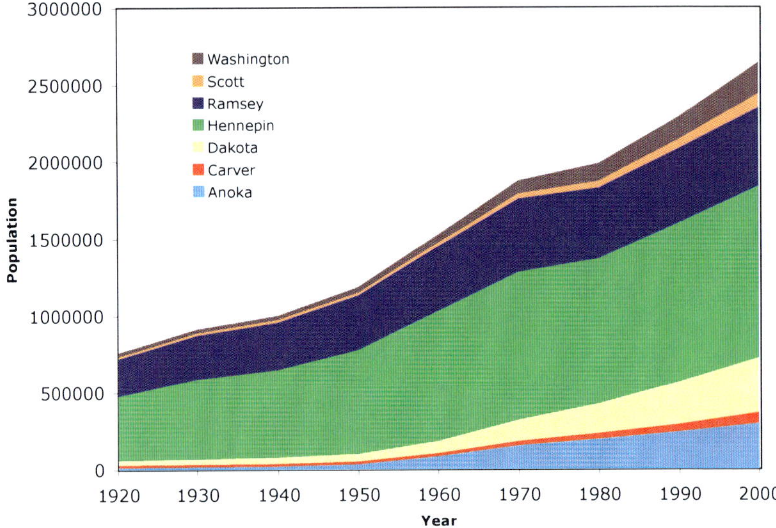

Fig. 7.2 Population growth by county in the Twin Cities during 1920-2000

7.6 Findings and concluding remarks

This chapter investigates the existence or absence of first mover advantages in the transportation sector using four empirical examples.

The case of London rails lends credibility to inherent first mover advantage in the deployment of rail stations. On both the surface rail and Underground networks, the early stations have more connections than later stations, and account for a larger share of ridership. A more rigorous analysis of station boardings and alightings suggests that the first mover advantage is closely associated with the spatio-temporal location of the stations and network connectivity.

Examining the global aviation system suggests there has been a significant persistence of airlines at hubs. Five of the six US network carriers retain at least one hub from their earliest days, and if considering both passenger and mail service, all six airlines are still serving at least one of those markets.

A port is certainly an important aspect of a city's growth. A port that grows early may retain some disproportionate advantage for a time while equilibrium is established, and by doing so, may provide advantages for other complementary aspects of manufacturing and trade helping to enmesh its position. The evidence however suggests that first mover advantages are quite weak in this sector.

Analyzing a database of road projects in the Twin Cities suggests more recently constructed highways carry more traffic. The analysis, however, may be misleading as it took into no consideration either the dynamics of roads (e.g., new roads may be built in place of old ones) or the development of places (e.g., faster growth in the suburb may counteract the locational advantages of early built roads in the urban core). An accurate treatment of first mover advantages would consider both transportation and land use in an evolutionary context. An empirical approach, however, is usually limited by the scarcity of extensive *ex post* data. Taking this, Chapter 11 will revisit the topic of first mover advantages using an *ex ante* modeling approach.

This part is dedicated to constructing analytical explanations of network growth. A series of *ex ante* models are introduced to interpret the spontaneous development of transportation networks from innovative perspectives, and are applied to investigating the spontaneous organization of large-scale network features such as hierarchy, topology, and sequence..

Chapter 8
Hierarchy

8.1 Introduction

Driving in the United States and other countries, one could make an educated guess about the relative importance of a road by looking at its route marker. See Figure 8.1 for a collection of route markers from different ranks of interstates, U.S. Highways, state highways, and county roads. Roads of higher ranks tend to be wider, faster, and carry more traffic. Taking Minnesota as an example, there are approximately 19,300 kilometers (12,000 miles) of Interstates and state highways (9% of the total road length) as of 2005, which account for about 60% of the total 87 billion annual vehicle-kilometers (54 billion vehicle-miles) traveled in this state (Minnesota Department of Transportation, 2005*a*).

Fig. 8.1 Examples of road markers

Transportation infrastructure networks possess system properties that are globally observed across mode, space, and time. As we can see from the above example, there is a hierarchy of transportation links, such that some links are more important than others. These properties (width, speed, capacity, among others) are traditionally thought to be the product of conscious design or planning. In other words, planners or policy-makers designate the formation of transportation networks and decide on their collective characteristics in a centralized, top-down manner. This view is not unrealistic in the sense that historically transportation infrastructure was often designed and provided at a central level by a governmental or private authority. This centralized view, however, neglects how individual travelers, property owners, private developers, and localities respond to changes in transportation infrastructure networks, and how their independent initiatives interact and find their way to subsequent transportation policies or plans in a bottom-up process.

In reality, the development of transportation networks is played out as the outcome of both centralized and decentralized decision-making processes. Since the decentralized side of the story is largely neglected in the literature, a main theme of this book is to construct models that explain the evolutionary growth of transportation networks from a more balanced perspective. To start with, this chapter will demonstrate the decentralized development of a transportation network by taking it to the extreme, i.e., assuming completely localized investment decisions on the network. Based on simple, myopic, and decentralized investment processes in a network, this study will explore if the network can spontaneously organize itself into a hierarchical order during network dynamics.

The self-organizing property of network development was originally studied in network science (see Section 2.5 for a comprehensive review). In the last decade, network scientists have found that many complex network systems such as the world-wide web and metabolic networks seem to spontaneously evolve into large-scale order, even based on simple behaviors of independent agents in the systems (Schelling, 1978; Krugman, 1996). These so-called "scale free" networks unexceptionally exhibit a hierarchical structure in which some nodes acting as "highly connected hubs" are more important than others. These studies, however, mainly dealt with a node-centric world where spatial constraints of making a connection between nodes are not significant.

Historical observation bears witness to the spontaneous organization of transportation networks as well. Chapters 4 and 5 revealed the statistically significant relationship between the development of the skyway and Interurban networks and the connect-choices made by local or regional developers in their attempts to maximize network accessibility. Chapter 6, on the other hand, revealed that the reciprocal development of land use and transportation spontaneously led to the finger-shaped distribution of residences surrounding the streetcar network in Minneapolis and St. Paul and other mid-western cities in the late nineteenth and early twentieth centuries.

In explaining the dynamics of transportation networks, Lam and Pochy (1993) and Lam (1995) proposed an active-walker model (AWM) to describe the dynamics of a landscape, in which walkers as agents moving on a landscape change the landscape according to some rule and update the landscape at every time step. Helbing et al. (1997) applied the active walker model to explaining the emergence of trails in urban green spaces shaped by pedestrian motion. AWM models, however, are focused on primitive travel behavior (e.g., walking) that shaped a featureless landscape. Yamins et al. (2003) presented a simulation of road growing dynamics on a land use lattice that generates global features such as beltways and star patterns observed in urban transportation infrastructure. Yerra and Levinson (2005) and Levinson and Yerra (2006) investigated the self-organization of surface transportation networks using a travel demand model coupled with decentralized revenue, cost, and investment models.

To a large extent replicating the work of Yerra and Levinson (2005), this chapter presents a network growth model based on simple, decentralized investment decisions made by local "agents" that represent individual links, and uses this idealized model to explore the hypothetical question that whether or not the hierarchical structure of transportation networks, rather than being designed by central authorities, is an emergent property of spontaneous organization.

8.2 Model

For the purpose of this study, a model called System of Network Growth (SONG) is developed to demonstrate the self-organization of surface transportation networks. Figure 8.2 outlines SONG's component models and their connections in a flowchart. Component models and their mathematical formulations are explained below.

8.2.1 Land use allocation

Following Yerra and Levinson (2005), the SONG model represents transportation and land use in a schematic two-layer structure. The transportation network is represented as a directed graph that connects nodes with directional links. Surrounding land use activities are distributed over a hypothetical space that is divided into a two-dimensional grid of land use cells. In this analysis, the model was executed over a 50×50 grid lattice of land use cells. On top of it is an evenly spaced 10 node by 10 node grid network. For the sake of a simple demonstration, it is assumed that each land use cell generates a pre-specified number of trips, which are assigned onto the network at its nearest node. The time cost of accessing the nearest network node from a land use cell is called the "access cost", which will be explained later.

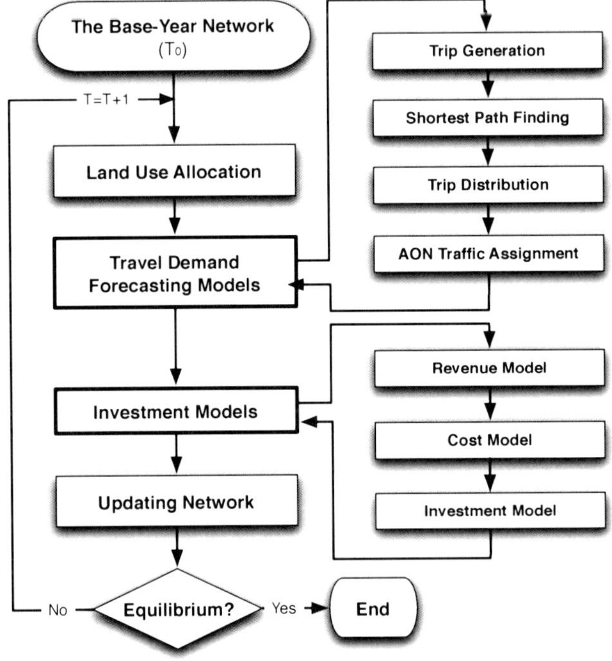

Fig. 8.2 Model framework of SONG: System of Network Growth

8.2.2 Travel demand models

The SONG model includes a travel demand model that translates exogenous land use data into traffic flows on a pre-existing network. A traditional travel demand forecasting model includes four steps to predict the traffic flows on a given road network: trip generation, trip distribution, mode split, and traffic assignment (Ortuzar and Willumsen, 2001). For simplicity, this analysis skips the mode split step by assuming a single abstract mode in the travel demand model. The steps of trip generation, shortest-path finding, trip distribution, and traffic assignment are followed for each time period to predict through traffic across individual links, discussed in turn.

In the *trip generation* step the total trips produced from and attracted to a network node are calculated by summing up trips produced in or destined for all the land use cells attached to this network node.

In the *shortest-path finding* step, a shortest path, i.e., the path with the lowest generalized travel cost, from each node to all other nodes in the network is calculated using Dijkstra's Algorithm (Chachra et al., 1979).[1] Let t_a^i represent the generalized

---

[1] It needs to be noted that Dijkstra's algorithm does not enumerate all possible shortest paths between two nodes. When there are multiple equal cost paths between an origin and destination,

cost traversing link a in Time Period i. This cost is calculated as the linear combination of time cost and monetary cost (in this study, toll represents all monetary costs spent on travel) as shown in Equation 8.1 below:

$$t_a^i = \eta^{l_a}/v_a^i + \tau(l_a)^{\rho_1}(f_a^i)^{\rho_2}(v_a^i)^{\rho_3}, \forall a \in \{A^i\} \tag{8.1}$$

Where η is time value, l_a is the length of link a, f_a^i and v_a^i are the average flow (volume) and average speed of link a at Time Period i. Coefficients τ, ρ_1, ρ_2, and ρ_3 denote toll rate, length coefficient, flow coefficient, and speed coefficient specified in the revenue model, respectively.

The bold curve in Figure 8.3 illustrates the shortest path from node R to all other nodes. The shortest path from R to S is denoted as P_{RS}, which represents the set of consecutive links along the shortest path from origin R to destination S. Suppose m land use cells r_1, r_2, and, r_m are attached to node R and n land use cells s_1, s_2, and s_n are attached to node S. The travel cost from origin R to destination S along the shortest path for iteration i can be calculated as:

$$t_{RS}^i = {}^1\!/\!_m \sum_{j=1}^m (\eta d_{r_j}/v_0) + \sum_a t_a^i \delta_{a,RS}^i + {}^1\!/\!_n \sum_{k=1}^n (\eta d_{s_k}/v_0) \tag{8.2}$$

The first part on the right side of the above equation calculates the average access cost walking from the land use cells attached to Node R to this node. The variable d_{r_j} represents the distance from a land use cell r_j to node R. The variable v_0 is a specified minimum speed, which can be interpreted as the walking speed for accessing the closest network nodes from the land use layer. Similarly, the third part calculates the average access cost from node S to the land use cells attached to it. The second part sums the generalized costs of the links in P_{RS} where $\delta_{a,RS}^i$ is a dummy variable which equals 1 if a link belongs to P_{RS} and 0 otherwise.

A trip table (Origin-Destination matrix) is computed using a doubly constrained *trip distribution* procedure (Hutchinson, 1974). The interaction between places assumes a gravity-type form:

$$N_{RS} = K_R K_S O_R D_S f(t_{RS}) \tag{8.3}$$

Where O_R indicates number of trips produced from origin R while D_S represents number of trips attracted to destination S. Coefficients K_R and K_S represent K- factors to be estimated in the procedure. In this analysis, The gravity function adopts a negative exponential form:

$$f(t_{RS}) = e^{-\theta t_{RS}} \tag{8.4}$$

Where θ is a global friction factor in trip distribution.

Dijkstra's algorithm chooses one based on the arbitrary numbering of nodes. This can cause issues when a network is either symmetric or can be mapped onto itself by permuting the node numbers. Following Levinson and Yerra (2006), this analysis addresses this issue by externally applying symmetric conditions, more specifically, by averaging link flows with their images from all possible symmetric axes. Please refer to Levinson and Yerra (2006) for more detailed explanations.

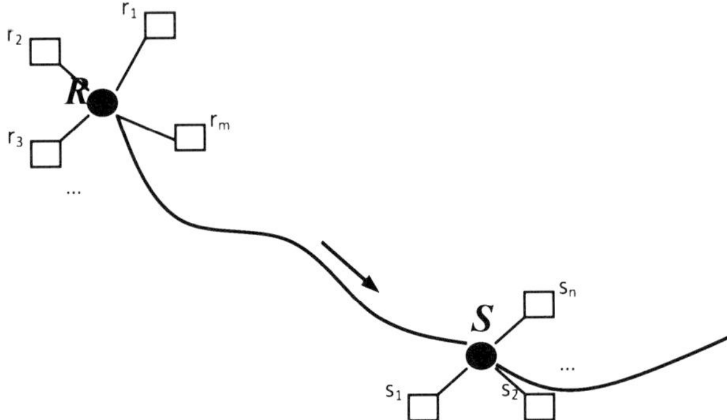

Fig. 8.3 Calculation of the generalized path cost

Since this analysis imposes no capacity constraint on a network, an all-or-nothing (AON) *traffic assignment* procedure is adopted, by which all the trips between a node and every other node in the network are assigned onto the least-cost path between the origin and destination.

8.2.3 Investment

Investment models represent completely localized investment decisions on a network. These decisions in terms of tolling, spending, and investing are abstracted in simple equation forms, also assuming autonomous links make myopic decisions without considering cooperating with others or saving for the future. To ensure two parallel and opposite one-way links a and b that connect two nodes are always maintained at the same conditions, we assume that a single agent operates both links as a whole. A *revenue model* calculates revenue collected on each individual link depending on speed, flow and length of this link. In each time period, the agent gathers revenues from the toll collected by both member links, computes the overall maintenance cost it spends on both links, and then decides to invest in its speed level.

Let f_a and f_b respectively represent the flows of link a and link b for iteration i. Earnings accruing to the link result from tolls that are collected by link agent ab as follows:

$$E_{ab} = \tau l_{ab}(f_a + f_b) \tag{8.5}$$

Where l_{ab} is the length of each link (given that link a and link b have the same length) while τ represents the specified toll rate.

A *cost model* calculates the cost to maintain links in their present usable conditions, which falls into two parts: the variable cost depends on link length, flow and speed, while the fixed cost is independent of flow and speed and only depends on link length. The overall maintenance cost spent by the agent operating links a and b for iteration i is calculated by adding up the fixed and variable costs spent by both links as:

$$C_{fixed,ab} = 2(l_{ab})^{\alpha_1} \tag{8.6}$$

$$C_{var,ab} = (l_{ab})^{\alpha_1}[(f_a)^{\alpha_2} + (f_b)^{\alpha_2}](v_{ab})^{\alpha_3} \tag{8.7}$$

Where α_1, α_2 and α_3 represent specified length, flow, and speed powers in the cost equation, respectively.

An *investment model* updates the service level of every link (surrogated by link speed) at the end of each time step depending on the revenue and maintenance costs. It is assumed each link agent myopically decides to spend all available revenue without saving for the next time step. If the revenue (net of fixed costs) gathered from links a and b exceeds the maintenance cost spent for iteration i, remaining revenue will be invested to improve the service level, that is, the running speed of links a and b. In contrast, if the revenue is insufficient to cover the cost, the running speed will be downgraded. The investment mechanism is designed in a way that, the greater is the revenue to cost ratio, the higher is the updated speed on a link. This myopic investment policy is formulated as:

$$v_{ab}^{i+1} = v_{ab}^{i} \left(\frac{\left(E_{ab}^{i} - C_{fixed,ab}^{i}\right)}{C_{var,ab}^{i}} \right)^{\beta} \tag{8.8}$$

Where v_{ab}^{i+1} and v_{ab}^{i} are respectively the speed levels of link agent ab for iteration $i+1$ and iteration i, while β is a specified speed improvement coefficient.

At the end of a time step, the link-based investment processes update the network, which is then fed into the next time step. One time period represents a hypothetical year as the day-to-day traffic on the network is predicted in the travel demand models and converted to yearly traffic for investment models. The process is repeated until an equilibrium state of link speeds is reached, or it is clear that no equilibrium can be reached.

8.3 Hypothesis and experiments

The main research question of this study is whether or not hierarchy is an emergent property of network dynamics. It is hypothesized that a transportation network will spontaneously evolve into a hierarchical order based on decentralized investment processes specified in the SONG model, regardless of its initial status.

To test this hypothesis, four experiments were conducted to evolve the network under different initial conditions. Table 8.1 lists model parameters whose values are specified to our best knowledge of economies of scale in transportation economics.

Table 8.1 Model parameters and their specified values

Parameter	Description	Value
v_0	Walking speed	0.01
τ	Toll rate	1
α_1	Length power	1
α_2	Flow power	0.75
α_3	Speed power	0.75
β	Speed improvement coefficient	1

Experiment 1 (the base case) assumes a uniform initial speed of 5 across individual links in the network.[2] It is also assumed that each land use cell generates and attracts the same number of 10 trips per day.[3]

SONG visualizes the spatial distribution of link speeds across the network at the end of each time period. The entire range of link speeds is divided into 5 intervals (0.01-5, 5-10, 10-15, 15-20, 20+). Each interval category is displayed using a unique combination of line thickness and color. The thicker is a link, the higher is the corresponding speed. Snapshots of the evolving network in Experiment 1 are presented in Figure 8.4. The left figure in Figure 8.4 displays the initial network; the middle displays the intermediate network after one iteration of network dynamics, and the right displays the resulting network after 10 iterations when equilibrium is reached. Because of the symmetric initial land uses and link speeds, the resulting networks from the base case all exhibit symmetric patterns. The stabilized network clearly exhibits the emergence of network hierarchy based on link-based investment processes. The fastest links form a ring in the center, while express links extend to the suburbs. The emergence of higher-hierarchy links in the center could be attributed to the boundary effect - while links on the border are only traversed by inward trips, links in the center are used by trips from every direction.

Experiment 2 is similar to the base case except for randomly distributing the initial link speeds between 1 and 5. Ten different sets of random numbers were tested, and snapshots of a typical solution are displayed in Figure 8.5. As can be seen, starting from a random state, the network still appeared somewhat chaotic at the end of the first time period, but it quickly organized itself into a stable, hierarchical order when it reached the equilibrium at the end of the 9th time period. Because of the asymmetric distribution of initial link speeds, the resulting networks are asymmet-

---

[2] As an abstract representation of network dynamics, the SONG model assumes no units associated with its model parameters.

[3] Central place theory (Christaller, 1933) suggests that places form hierarchies over space. Intuitively, hierarchies of places would lead to a hierarchical network with more important links serving places with more desired activities. To eliminate the effects of place hierarchy and focus on the inherent property of transportation systems, this analysis assumes either uniformly or randomly distributed land uses over space.

ric too. Faster links, however, still emerged in the central area connecting with each other into a web.

Experiment 3 is similar to the base case except for randomly distributing the initial land use between 0 and 20 (while the total number of trips generated or attracted from land use activities is controlled). As can be seen in Figure 8.6, the random distribution of land uses resulted in a different pattern of link speeds which is both asymmetric and hierarchical.

In Experiment 4, initial speeds are randomly distributed as they are in Experiment 2, and land uses are randomly distributed as in Experiment 3. Despite different initial land use distributions, the network evolved into a pattern which is similar to the results in Experiment 2, with faster links in the center forming a web-like pattern (Figure 8.7).

Taken together with the above observations, it is clear that the tested transportation network evolved into different patterns depending on the distribution of initial speeds and land uses. Irrespective of the initial conditions, however, the network always organized itself into a hierarchical structure with some links becoming faster and carrying more traffic while others become slower and less used, based on completely decentralized investment decisions. This supports our hypothesis that transportation networks exhibit spontaneous organization during development; and hierarchy is one of the emergent properties of network dynamics.

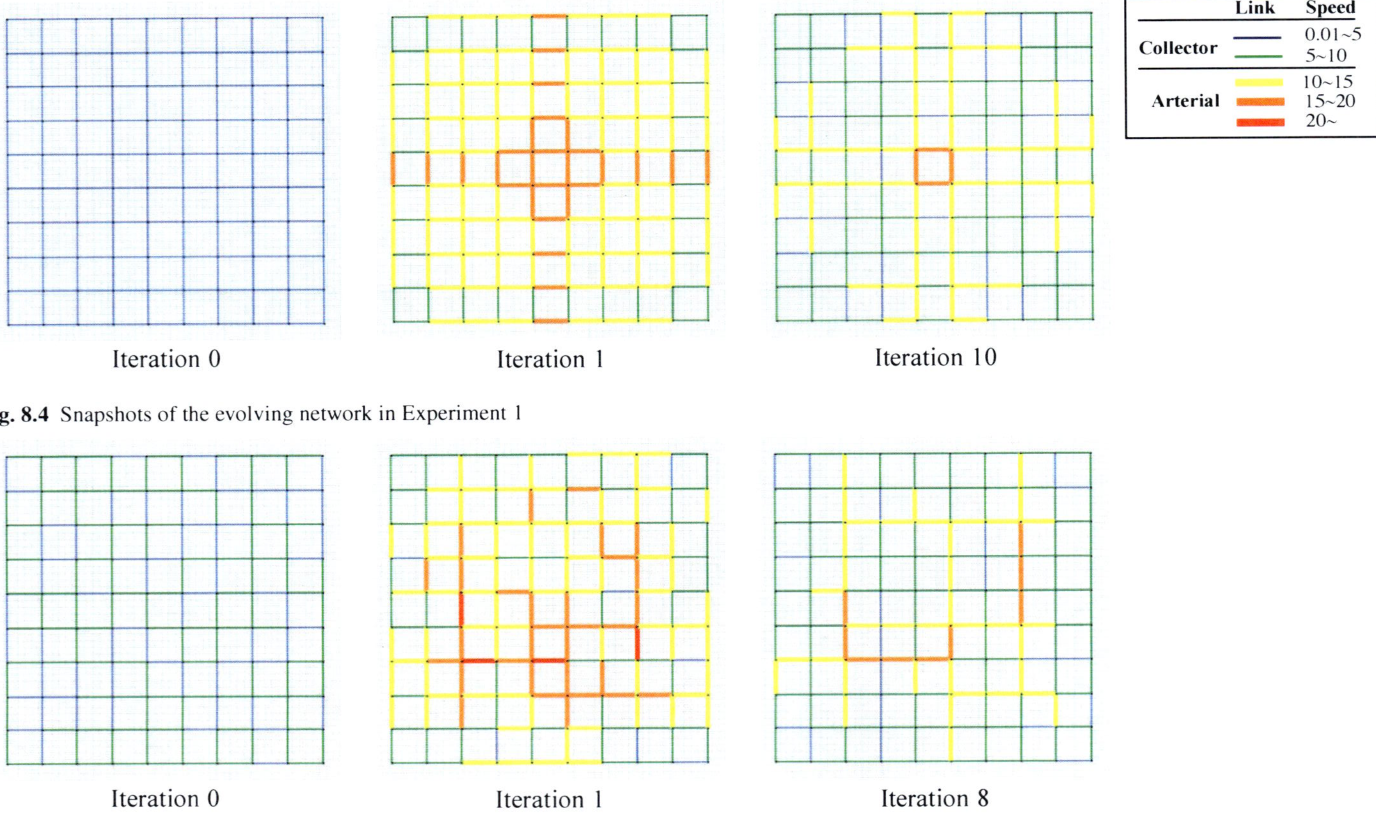

Fig. 8.4 Snapshots of the evolving network in Experiment 1

Fig. 8.5 Snapshots of the evolving network in Experiment 2

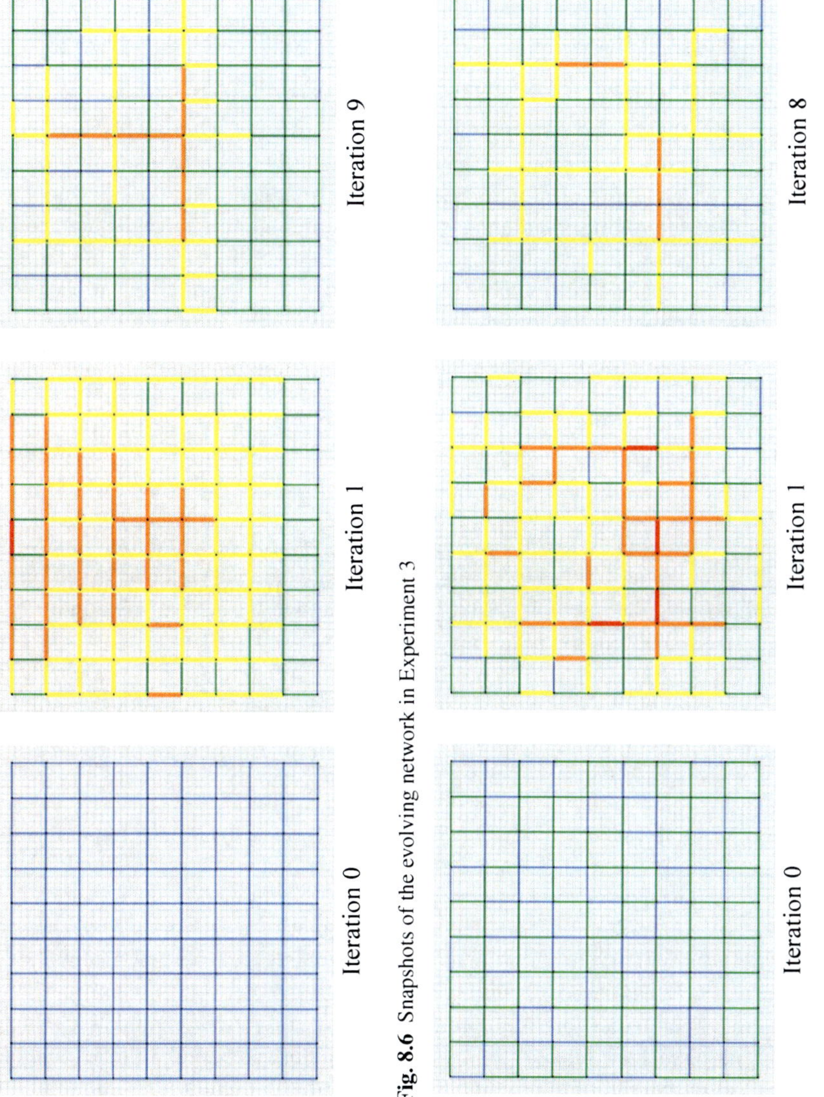

Fig. 8.6 Snapshots of the evolving network in Experiment 3

Fig. 8.7 Snapshots of the evolving network in Experiment 4

8.4 Findings and concluding remarks

This chapter to a large extent replicates Yerra and Levinson (2005) in demonstrating the spontaneous development of transportation networks based on decentralized investment processes. The SONG model was constructed to evolve a transportation network until an equilibrium state is reached. Experiments conducted on an idealized grid network over a hypothetical space support our hypothesis that based on simple, myopic investment decisions made by independent proxies of individual links, a transportation network can spontaneously evolve into a stable hierarchical order regardless of its initial state. The findings from this analysis suggest that rather than being designed or planned in a top-down manner, the development of transportation networks could be spontaneously driven by decentralized, bottom-up decisions.

The SONG model is the first *ex ante* model that this book introduces to interpret the evolutionary growth of transportation networks. Due to the complexity our model involves, compromises have to be made between realism and feasibility. For the purpose of a clear demonstration of spontaneous network development, the SONG model is simplified in many ways. Major assumptions made in the model are discussed below.

First, the SONG model includes completely localized investment processes by which simple, myopic decisions are made by independent agents representing individual links. Historical observation in Chapters 4-6, however, suggests that the provision and operation of transportation networks lie somewhere between extremes of centralization and autonomy. Public or private stakeholders at a central or local level can come into play under their independent interests, and they can interact and compromise in a way that investment decisions on transportation networks are eventually made to their collective good. Chapter 13 and 14 will analyze the benefits and costs associated with investment decisions made at a central versus local level. Chapter 15 will examine investment decision rules developed based on surveys or empirical data, which are more representative of the realistic conditions in metropolitan transportation planning.

Second, the SONG model, considering only investment decisions that either contract or expand links on a pre-existing network, cannot represent the structural change of a network over time and space. The following chapters will present models that enable a variable network topology through degeneration or incremental connections.

Third, this analysis examines only hierarchy as one of the emergent properties of network dynamics, and the identification of hierarchical order is confined to visual observation. The next chapter will develop quantitative methods that capture a range of hierarchical, topological, and geometric attributes of a network. These measures will prove instrumental to trace the temporal change of transportation networks.

Fourth, the SONG model is only a demonstration of network dynamics. The values of model parameters are pre-specified without calibration or validation. Chapter 9 will demonstrate the feasibility of validating an abstract model to reveal the basic topological properties of a network. Chapter 15 will present a fully calibrated network growth model using the empirical data in a major metropolitan region. Chapter 11 and Chapter 12, on the other hand, will employ sensitivity analysis to examine the variations of model outputs to specific parameters.

Fifth, the SONG model eliminates congestion effects by assuming infinite link capacity in traffic assignment.[4] This is enhanced in Chapters 12, 14 and 15 by introducing capacity constraints on links and Stochastic User Equilibrium traffic assignment.

Last but not least, the coupled development of transportation and land uses represents a two-way process by which transportation and land use affect each other. The SONG model neglects the effects of land use development by specifying exogenous land use inputs. The co-development of transportation and land use will be studied in Chapters 11 and 12.

---

[4] It is worth noting, though, while congestion is neglected in the travel demand model, it is indirectly accounted for in the investment models, as congested links generate higher revenue, and therefore get more investment.

Chapter 9
Topology

9.1 Introduction

Originally conceived as a means of diverting through traffic away from congested central-city areas, circumferential limited-access highways (often referred to as beltways, ring roads, or loops) have become integral parts of the intra-metropolitan highway system. Many cities and metropolitan areas have their distinguishing beltways. To name just a few, the Berliner Ring, the London Orbital, and the Capital Beltway. Figure 9.1 displays the sketch highway networks of eight beltway cities in America during 1976-1980.

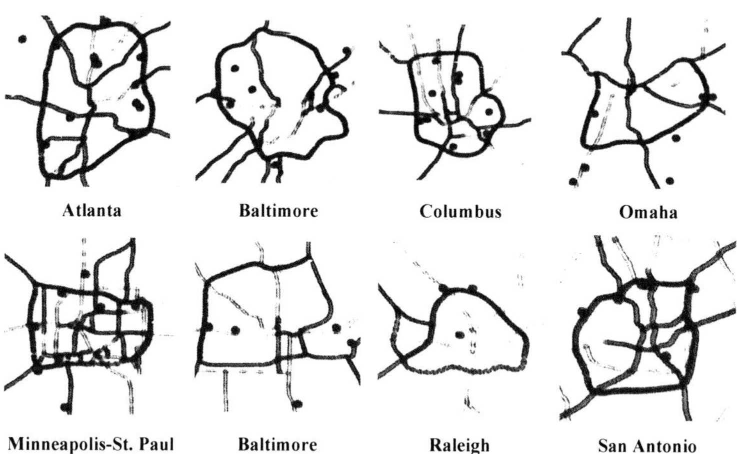

Fig. 9.1 The 1976-1980 highway networks of eight beltway cities in America. Bold lines represent highways and dots represent regional shopping centers. Data source: Payne-Maxie Consultants (1980).

As evident from the beltway example, transportation networks are characterized by their geometric patterns. Certain patterns like hubs-and-spokes are typical in railway and airline systems, while grids, beltways and cul-de-sacs can be observed in many urban road networks. Extending the efforts in Chapter 8, this chapter studies the spontaneous organization of network topologies from an evolutionary view.

A variable network topology is enabled by a new model of network dynamics in a spontaneous process of network degeneration. The essential idea of network degeneration is that the temporal development of a surface transportation network can be viewed as a degeneration process: starting from an underdeveloped area where all point-to-point paths can be used, those paths which are more valuable are reinforced while less used ones shrink and are finally abandoned. Taking road infrastructure as an example, while dirt trails and turnpikes built in the early stage of surface transportation disappeared on less used routes (such as those connecting villages to villages), those on valuable routes (such as those connecting towns to villages and to other towns) survived and were replaced by paved roads, some of which may be further upgraded into into arterial, highways or freeways. If our focus were on paved roads rather than the whole spectrum of facility types, we would observe a network of paved roads that gets increasingly connected through time until they are replaced by infrastructure constructed with a newer technology. In this sense, the degeneration process represents the same "growth" process of transportation networks we have observed in reality. Based on this idea, we develop a network growth model that incorporates decentralized investment processes on individual links, by which much used links continuously get reinforced while less used ones shrink. Furthermore, by shuttering the weakest link in a network at the end of each time step, the models enables a variable network topology in a discrete process.

Analysis of network topologies requires more than visual recognition. Another focus of this chapter is to introduce quantifiable indicators which would help abstract the properties of complicated network structures and interpret the structural transformation of these networks in a temporal context. More importantly, this research will demonstrate the feasibility of validating an abstract model to reveal the basic topological properties of a network.

The rest of the chapter takes the following form: the model is first constructed. In the next sections, spatial measures of network topologies are introduced and then the model validated against historical data extracted from the Indiana interurban network during its decline phase between 1917-1941. The model is then applied to an array of idealized networks in simulation experiments to examine the topological evolution of surface transportation networks. The conclusion summarizes our findings and indicates future directions.

9.2 Model

This section constructs System Of Ultra-connected Network Degeneration (SOUND) to represent the evolutionary growth of transportation networks using a degeneration process. Alternatively, the model can be thought of as simulating a mature system where all links have been built, and some links are abandoned while others are improved. As a whole, the evolution of a network is represented as an iterative process over discrete time (simulation) periods and each period implements topological changes in a sequential process that includes four consecutive components: land use allocation, travel demand dynamics, investment, and disinvestment. While the disinvestment process may be terminated when specified stopping rules are met, the process will be repeated until an equilibrated state is reached. Figure 9.2 illustrates the procedure of the SOUND model in a flowchart. As can be seen, the SOUND model implements the same process as the SONG model presented in the preceding chapter, except for an additional disinvestment step following the investment models, which will be explained below. The explanations of the land use allocation, travel demand dynamics, and investment models can be found in Chapter 8.

The disinvestment process eliminates the least used link(s) from the collection of existing links for each time period. In this case, the usage of a link[1] is represented by the volume of its through traffic. The disinvestment criterion specifies how the weakest links are selected and ensures a minimal number of links to be eliminated for each time period. For an asymmetric network, one and only one link agent that operates at the lowest flow level will be killed at a time among removable links. A removable link agent is one which, by removing both its member links, does not disconnect the network. For a symmetric network with symmetric demand, all the links on symmetric positions operate at the same speed and flow level. Thus once one link is selected to be killed, all its symmetric counterparts will be selected and removed automatically as well. As the disinvestment process removes links from a network, a node may become disconnected when all its connections are removed. In the next time period, a land use cell will then re-allocate its trips to the second nearest node if the node to which it is currently attached is removed during the degeneration process. The distance from each land use cell to its nearest node is re-calculated as well. The "weakest-link" assumption aims to capture the essence of realistic decision-making: myopic and locally optimal made in a sequential, path-dependent process of transportation development.

9.3 Measurement

The long-standing interest in the spatial structure of transportation networks has been driven by the profound impact of network structure on urban mobility, as well as its critical role in the formation of travelers' behavior, residential patterns and city morphologies. Consequently, a wide range of spatial measures has been developed in the literature to quantify important collective features of transportation networks such as connectivity, heterogeneity, density, and geometric connection patterns. The

---

[1] Consistent with the SONG model, a link contains two directional arcs on opposite directions.

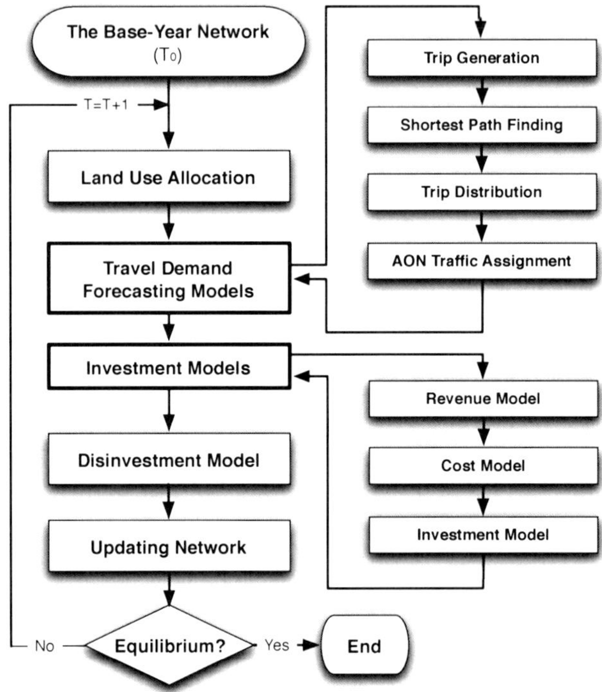

Fig. 9.2 Model framework of SOUND: System Of Ultra-connected Network Degeneration

section introduces some important measures that will be used in this chapter and throughout the rest of this book.

9.3.1 Connectivity

Early work on measuring the structure of transportation networks dates back to the 1960s, when geographers and transportation researchers focused almost exclusively on graph-theoretic measures in network analysis (Garrison, 1960; Garrison and Marble, 1962; Kansky, 1963; Haggett and Chorley, 1969). Four elementary graph-theoretic indices include Cyclomatic number, Alpha index, Beta Index, and Gamma Index.[2] The cyclomatic number indicates the number of circuits in a network. The alpha index is the ratio between the actual number of circuits in the network and the maximum number of circuits; the beta index is the ratio between the number

---

[2] It should be noted that the Alpha index as a statistic here is different from the model parameters α_1-α_3 outlined in Table 8.1, and the Beta Index is different from the speed improvement coefficient β.

of links and the number of nodes; the gamma index compares the actual number of links with the maximum number of possible links in the network. These three indices can estimate the multiplicity of links in a network (values for the three indices range from 0 to 1 with a higher value indicating a more connected network), and can also form some useful common yardsticks for comparison between networks. The gamma index (γ), in particular, is defined and calculated for this research as follows:

$$\gamma = \frac{e}{3(v-2)} \tag{9.1}$$

Where e is the number of edges (directional links) and v is the number of vertices (nodes).

9.3.2 Density

Another simple yet useful graph-based measure is network density (D), which measures the length of links per unit of surface. The higher it is, the more a network is developed. The density of a network is measured by the length of examined links (L) divided by the area of the territory (B).

$$D = \frac{L}{B} \tag{9.2}$$

9.3.3 Heterogeneity

In recent years, network research has shifted its focus to large-scale statistical properties of complex networks such as heterogeneity (hierarchy) and nodal connectivity (Albert et al., 1999; Barabási, 2002; Barabási and Bonabeau, 2003; Newman, 2003). In a scale-free network, nodal connectivity is indicated by the degree of that node (number of connections) (Newman, 2003) and using each node's degree as a proxy for its importance, the heterogeneity of a complex network can be further statistically quantified (Sole and Valverde, 2004; Trusina et al., 2004). Without taking into account link properties, however, few of these node-centric measures can be used for transportation networks.

This research introduces entropy as an indicator of the heterogeneity of link attributes in a transportation network. The concept of entropy was initially proposed by Shannon (1948) in his landmark paper "A Mathematical Theory of Communication" to measure information uncertainty, and has been widely introduced in recent years to measure the heterogeneity of complex network systems (Balch, 2000; Sole and Valverde, 2004). If individual links of a transportation network are considered as a collection of agents, they can be grouped into subsets based on different link attributes such as speed, functional type, traffic volume, or level of service. The proportion of each subset is calculated as the frequency of links in this subset over the total number of links, and then proportions are aggregated into an entropy measure that indicates the differentiation of the system. For example, the entropy measure of heterogeneity (H) with regard to link speeds is defined as:

$$H = -\sum_{k=1}^{\infty} p_k \log_2(p_k) \qquad (9.3)$$

Where p_k is the proportion of links on the kth level with regard to the total number of existing links; speeds are organized into a histogram and links whose speeds fall into the range k-1 to k are categorized into the kth level.

The Gini index (G) is another important indicator of system heterogeneity. Chatterjee (2003) elaborates the computation of the Gini index based on the Lorenz Curve. Following Lämmer et al. (2006), the Gini index can be used to reflect the concentration of traffic along links in a network. The importance of a links can be characterized by the number of vehicles or passenger that pass through it within some time interval. Accounting for the effect of heterogeneous link lengths, the actual use of a link is measured by the daily vehicle kilometer travel (VKT) or passenger kilometer travel (PKT) that occurs on a link. As the travel demand model allows the prediction of traffic volume on each link, the Gini index of VKT or PKT distribution on the network is approximated in a discrete form as follows:

$$G = 1 - \sum_{k=1}^{K} (X_k - X_{k-1})(Y_k + Y_{k+1}) \qquad (9.4)$$

Where X_k represents the cumulative portion of links for $k = 0, 1, ..., K$ while Y_k represents the cumulative portion of total VKT or PKT. Links are sorted in ascending order according to VKT or PKT that occurs on individual links.

The Gini index can be used to quantify the spatial concentration of network infrastructure as well. If Y_k in Equation 9.4 represents the cumulative portion of total link capacities rather than VKT, the index can instead measure how evenly infrastructure capacities are distributed over a network. Levinson and Zhang (2004) and Levinson and Zhang (2006) used the Gini index of delay in the study of equity on ramp meters. Similarly, the Gini index can also be used as a measure of spatial agglomeration of urban space, indicating how evenly land use activities (such as population and employment) are distributed over a set of land blocks (zones). Please refer to Krugman (1996); Chatterjee (2003) for the calculation and application of the Gini index in urban studies.

In analogy with kinematics, measures of the moment of inertia (I) and the equivalent radius (r) are introduced in this book to reflect the spatial clustering patterns of land use and network infrastructure. For example, the moment of inertia for the distribution of jobs in an urban area is computed as:

$$I = \sum_{k=1}^{n} J_k d_k^2 \qquad (9.5)$$

where J_k represents the employment of Zone k while d_j is the distance between the centroid of this zone and the center of the hypothetical metropolitan area.

The equivalent radius is then computed as:

$$r = \sqrt{1 \bigg/ \sum_k J_k} \qquad (9.6)$$

The equivalent radius essentially reflects how far away employment is distributed from the urban core. A radius of zero indicates all employment clusters in the center of the region while a larger radius indicates employment is located farther away from the center. Similarly the equivalent radius can also be computed for the spatial distribution of any other type of land use activities as well as network infrastructure.

9.3.4 Connection patterns

In order to quantify typical geometric patterns inherent in transportation networks, we developed in a previous research (Xie and Levinson, 2007a) a graph-theoretic algorithm to identify pre-defined connection patterns in a given network, and proposed quantitative measures to evaluate the relative significance of each connection type. The significance of predefined structural elements including ring, web, circuit, and branch is defined and evaluated in the following equations:

$$\phi_{ring} = \frac{\sum_i (l_i \delta_i^{ring})}{\sum_i l_i} \qquad (9.7)$$

where l_i is the length of an individual link i; is equal to 1 when a link belongs to a ring. Similarly,

$$\phi_{web} = \frac{\sum_i (l_i \delta_i^{web})}{\sum_i l_i} \qquad (9.8)$$

Note that if a link is located on one and only one circuit, it belongs to a ring; if it is located on more than one circuit, it belongs to a web. If a link belongs to a web or ring, it is defined as a circuit link; otherwise it is defined as a branch link. Therefore,

$$\phi_{circuit} = \phi_{ring} + \phi_{web} \qquad (9.9)$$

$$\phi_{tree} = 1 - \phi_{ring} - \phi_{web} \qquad (9.10)$$

9.4 Model validation

Previous studies have revealed the similarity between topological patterns generated in their simulation models and those observed in reality. Visual similarity, however, is not sufficient to support the validity of these models. This section extracts empirical data from a historical transportation network that actually experienced decline, and validates our model by comparing historical observations with simulation results based on non-parametric statistical analyses.

Among those surface transport modes that have experienced decline (canals, turnpikes, passenger rails, streetcars, interurbans, etc.), the interurbans in North America have left us probably the best archived historical records. Chapter 5 examined the expansion phase (1887-1916) of the Interurban in Indiana, America's second largest network of interurbans of the day. The network had its first line in 1887, started to decline from 1917, and completely disappeared in 1941. As shown in Figure 9.3(a), the interurban network of Indiana in 1916 represents the network in its full shape.[3] The decline phase of the network (1917-1941) is taken as a case to validate our model as follows.

Starting from the 1916 interurban network, a simplified travel demand model[4] is adopted to estimate the traffic flows on individual links, and the link operated on the lowest simulated volume is removed for each time period based on the "weakest-link" heuristic until the whole network disappears (i.e. in this case no stopping rule is imposed). A Spearman's rank-order correlation test (Higgins, 2003) is then taken to correlate the "predicted" sequence of link closure in simulation with the actual sequence of link closure (in which links are ranked according to the actual years of closure extracted from historic records). Spearman's rank-order correlation test assesses how well an arbitrary monotonic function describes the relationship between two variables, without making any assumptions about the frequency distribution of the variables.

---

[3] For the purpose of this study, the network of Indiana is separated from those other states basically along the border line. Please refer to Chapter 5 for more explanations on the development of the "complete" network in 1916.

[4] Please refer to Chapter 5 for the details of this travel demand model.

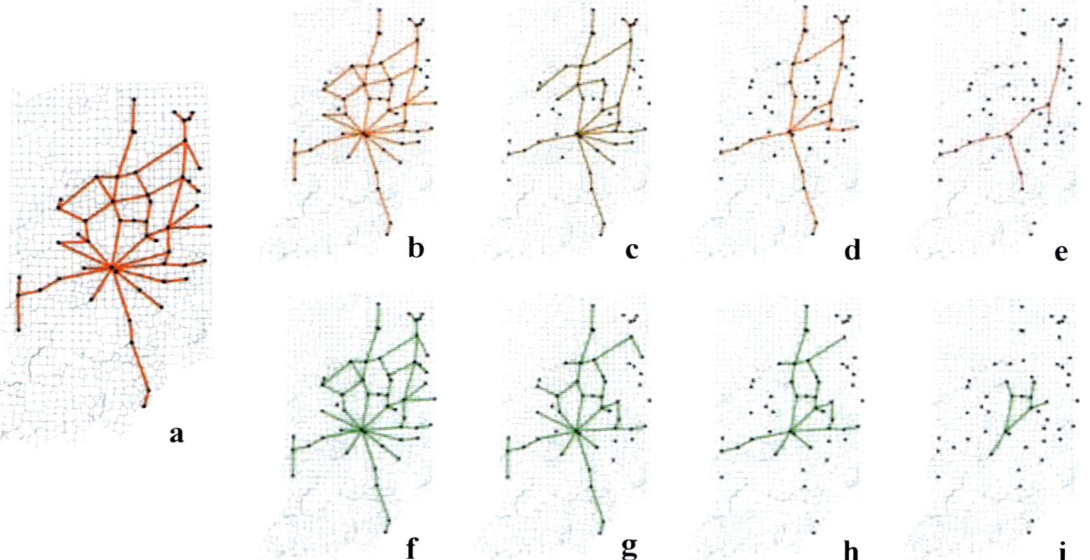

Fig. 9.3 Snapshots of the interurban network in observation versus in simulation during dynamics. Starting from the "complete" interurban network in (a) year 1916, the observed networks in (b) year 1930 (4 links closed), (c) year 1932 (21 links closed), (d) year 1935 (34 links closed), and (e) year 1940 (52 links closed), are respectively compared to the simulated ones in (f) iteration 5 (4 links closed), (g) iteration 22 (21 links closed), (h) iteration 35 (34 links closed), and (i) iteration 53 (52 links closed) with the same number of links remaining. The numbers in parenthesis indicate the cumulative number of links closed between the first year of decline and the year of examination.

According to the rank-order correlation test, the correlation coefficient is equal to 0.287. The *z-test* of correlation significance scores -2.245 with the *p-value* equal to 0.024, suggesting the two sequences are positively correlated at a 95 percent significance level. The correlation test implies that even based on a simplistic rule that closes links with the least traffic first, the "weakest-link" heuristic reproduced the sequence of link closure during the decline phase of the Indiana interurban network to a significant extent. Additional runs with different decay rates in the travel demand models show that the correlation is robust over the change of the decay factor and is even more prominent with a smaller decay rate.[5]

Figure 9.3(b)-(i) present the snapshots of the interurban network during the decline process in simulation versus in observation. To better illustrate, the topological change of the network is depicted in Figure 9.4 as the fluctuations of proposed topological measures both from simulation and from observation (the measure of circuitness is not included but treeness provides a complement measure). As can be seen in Figure 9.4(c), the measures of ringness in simulation indicate the emergence and collapse of a ring in the network, which is exactly what we observed in reality, although it is several iterations lagged as compared to reality. The series of each topological measure in simulation are compared to their counterparts in observation using the Two-sample Wilcoxon rank-sum (Mann-Whitney) test, which is a non-parametric significance test assessing if two independent samples come from the same population (Higgins, 2003). Viewing the measures as time-series data, this test essentially assesses if two series of data fluctuate over time following the same trend. The statistical results for each topological measure are also presented. Provided that the null hypothesis of the rank-sum test is that the two samples are drawn from a single population, none of the presented *z-values* provide sufficient evidence to reject the null hypothesis. Thus we can conclude that the computed attributes in simulation approximate the actual topological attributes of the network over time.

It is important to point out that, although empirical evidence from Indiana Interurbans lends credibility to our model, it does not necessarily indicate this model can be universally applied to other transportation networks with different geographic scales, of different modes, or in different regions without re-calibration or modifications.

---

[5] The decay factor in the trip distribution model indicates the rate of decline of the interaction between places across the interurban network. To test the sensitivity of the statistical results on the decay factor, model validation was re-executed with two different values of 0.005 and 0.02, and the Spearman correlation coefficient was re-calculated. With a lower rate of decline of the interaction (0.005), the Spearman correlation test results in a more significant correlation (the correlation coefficient equals 0.429 with a *p-value* of 0.0006), suggesting that under a higher degree of interaction, the "weakest link" heuristics performs even better in predicting the sequence of link abandonment. With a higher decline rate of 0.02, on the other hand, the Spearman correlation coefficient, though still with the positive sign, is much smaller and statistically not significant (the correlation coefficient equals 0.125 with a *p-value* of 0.289). This could be explained by the fact that as the decay factor increases, people become reluctant to travel farther. Provided that the interurban lines attract mainly intercity travel, the travel demand on the network may be underestimated, especially on strategic routes that connect big cities. This accordingly undermined the predictive performance of the "weakest link" heuristics.

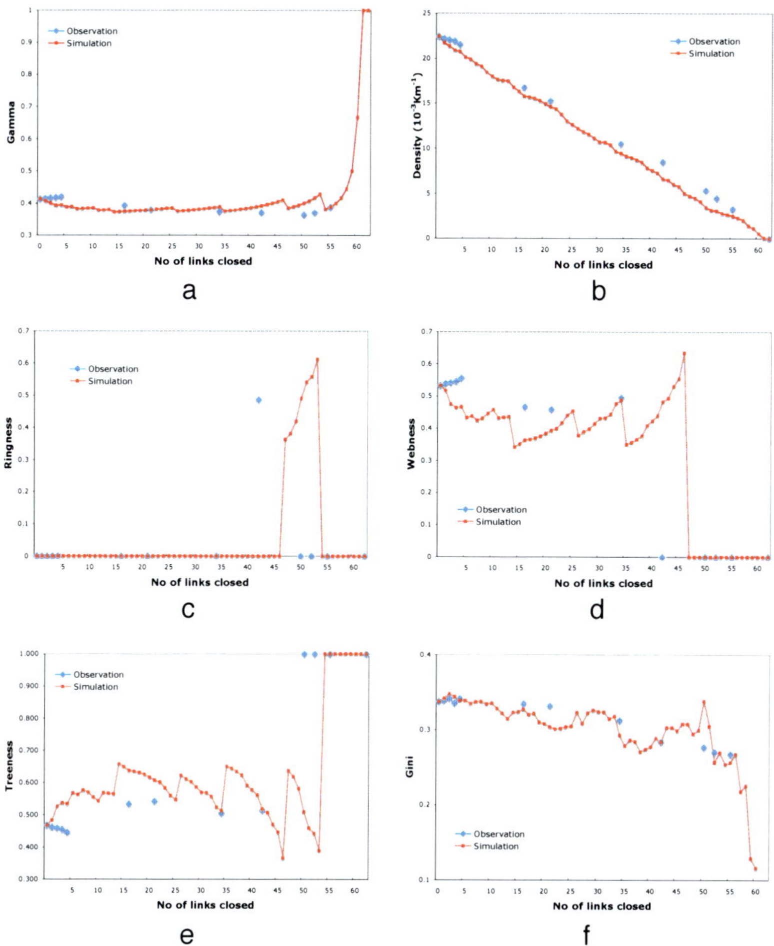

Fig. 9.4 Topological measures in simulation versus in observation are calculated from the dimensions of (a) Connectivity (z=0.014; *p-value*=0.989);(b) Density (z=-1.262; *p-value*=0.207); (c) Ringness(z=-0.375; *p-value*=0.707);(d) Webness (z=1.352; *p-value*=0.176); (e) Treeness (z=1.352; *p-value*=0.176); (f) Gini (z=1.347; *p-value*=0.178); In parentheses are the results of the Mann-Whitney test comparing the series of measures in simulation versus that in observation.

9.5 Simulation experiments

Now that the SOUND model is validated using the Indiana interurban data, it is applied with a range of assumptions on network structures and initial speeds to exploring (a) the temporal change of topological attributes emerging from different idealized network structures (b) typical topological properties that arise during the spontaneous evolution of networks, and (c) the sensitivity of emergent patterns to initial network conditions and to model parameters. A different stopping rule of minimum connection is used for the simulation experiments. To ensure the connectivity of the network, an existing node must connect to at least another existing node, and neither isolated nodes nor sub-networks are allowed during the disinvestment process. In the idealized ultra-connected networks developed for this study, a secondary node will be removed if all links connected to it are removed, while a primary node is not removable and it must be connected by at least one link.

9.5.1 Idealized network structures

Most cities have rather complex geometries of surface transport infrastructure. But from most origins in a city, one can travel locally in only four directions on its surface transportation network. In this sense, a surface transportation network has properties similar to a rectangular grid, in which travelers can only make a turn of $90°$ or a multiple of $90°$ from each node. The more directions which one can travel from any point, the shorter will be the average travel distance between points; at the same time the network will be more redundantly connected (Newell, 1980).

Aside from a grid network that simulates an actual urban road network, two ultra-connected networks are also developed, respectively referred to as the complete network and the hexagon network, which represent undeveloped networks where more directions are available moving from point to point.

A network developed by directly connecting every pair of two nodes among an original node set, which come from the intersection nodes of a grid network, is referred to as a complete network in this study. When links intersect, a new secondary node is created, and the longer link is replaced by shorter links that ultimately connect the same original nodes. A complete network does not necessarily directly connect each secondary node to an original node or another secondary node. Where links overlap, the longer link is eliminated. Note that theoretically the turning directions included in a complete network range from $0°$ to $360°$, depending on the size of the original grid network.

A hexagon network is also developed based on a grid network, in which the included angle of two intersecting links can be $30°$ or its multiples. Note that in the hexagon network, the primary nodes with 12 links connected are scattered on a lattice of equilateral triangles, and the secondary nodes connected to a primary node form a hexagon. This type of network has the same topology as the hexagonal landscape developed according to the transportation principle of Central Place Theory (Christaller, 1933; King, 1985), where smaller places are always located on the major transportation routes between the nearest larger places.

Since all these idealized networks are based on a square grid, their size can be indicated by the number of nodes along each side of the original square grid. For example, a 3×3 complete network is a network developed on a 3×3 grid. With the extent of a grid network fixed, its size also determines the average length of links and the spacing of parallel roads in the network. These networks are developed to represent transportation networks with geographical constraints, which as disclosed by Csányi and Szendrői (2004) are subject to the following scaling law:

$$U_S(w) \sim w^m \tag{9.11}$$

Where $U_S(w)$ denotes the size of neighborhood of node S within radius w, i.e., the number of nodes which can be reached from S in at most w steps. The scaling exponent m is defined by Gastner and Newman (2006) as the effective dimension of a network. In an infinite network it is calculated as:

$$m = \lim_{w \to \infty} \frac{\log U_S(w)}{\log w} \tag{9.12}$$

As transportation networks are finite, Lämmer et al. (2006) approximated the calculation by computing the average neighborhood size from a node in the network, plotting it over radius in double logarithmic scale. The slope of the curve at its inflection point gives the lower bound for an estimate of the dimension.

Figure 9.5 displays the estimated dimension of each idealized network, as well as that of the Indiana Interurban network in 1916 for comparison. As can be seen, the idealized networks demonstrate scaling properties that are similar to the Interurban network but distinct from scale-free networks, that is, the estimated dimensions for all the networks are strictly between 1 and 2, representing networks on the two-dimension surface of the Earth with limited sizes.

9.5.2 Simulation experiments

The model is run on different initial specifications in a series of six experiments listed in Table 9.1.[6] The experiments adopt the same set of parameter values as in the SONG model (Table 8.1), while the sensitivity of model outputs to these parameters will be further examined in Section 9.5.4. The results are visualized by displaying the pattern of link speed distribution across a network at the end of each time step. Following the preceding chapter, links are classified into five levels according to their speed values (0.01-5, 5-10, 10-15, 15-20, 20+), with different levels of links displayed with different colors and thickness.

---

[6] "Random" speeds represent specified initial link speeds randomly distributed from 1 to 10, "Uniform" speeds indicate initial link speeds equal to 5, and "Uniform land use" indicates the specified distribution of land use activities which produce 10 trips and attract 10 trips per day per cell.

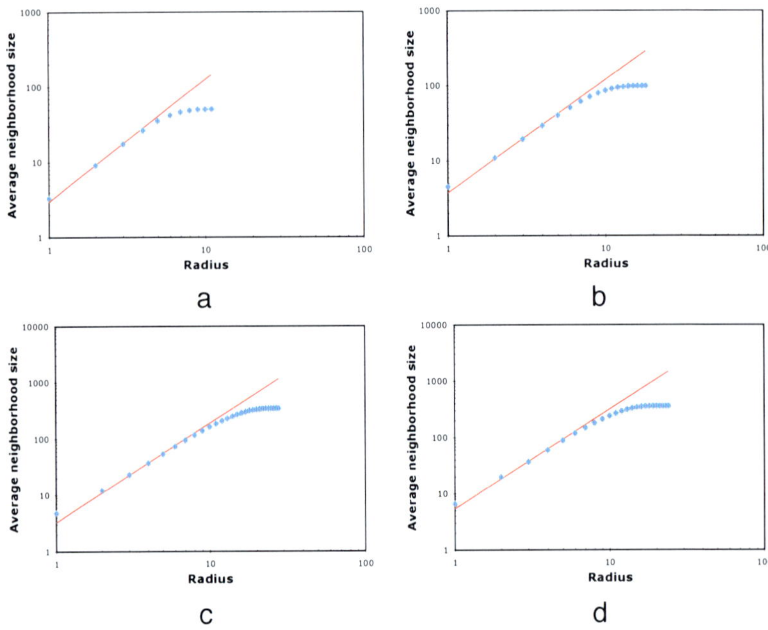

Fig. 9.5 Estimated effective dimension of (a) the 1916 Indiana Interurban network ($d = 1.63$); (b) the 10×10 Grid Network ($d = 1.50$); (c) the 4×4 Complete Network ($d = 1.76$); and (d) the 15×15 Hexagonal Network ($d = 1.77$)

Table 9.1 Specifications of experiments

No.	Initial conditions			Iterations
	Network	Link speeds	Land use	
1	10×10 grid	Random	Uniform	200
2	10×10 grid	Uniform	–	50
3	4×4 complete	Random	–	800
4	4×4 complete	Uniform	–	120
5	15×15 hexagon	Random	–	1000
6	15×15 hexagon	Uniform	–	300

A square 50-cell by 50-cell land use layer is adopted for all the experiments while three different network structures are tested: the 10×10 grid network, the 4×4 complete network, and the 15×15 hexagon network. For simplicity, a uniform distribution is specified for land uses under which each land use cell generates and attracts 10 trips a day. The initial speeds of links are also pre-specified as exogenous inputs and they determine the symmetry of a network. A network with a symmetric base structure, uniform land uses, and uniform link speeds evolves symmetrically in its topology, and thus it is called a symmetric network. Experiments 2, 4, and 6 were implemented on symmetric networks with uniform initial speeds and a symmetrical algorithm is included to ensure that the network evolved symmetrically; on the other hand, a network with random initial link speeds evolves asymmetrically and it is called an asymmetric network. Experiments 1, 3, and 5 were implemented on asymmetric networks with random initial speeds. Experiment 3 was repeated three times with different sets of random initial speeds, labeled as 3a, 3b, and 3c, respectively. Since all the experiments reached a stable equilibrium in the network growth model before the 20th iteration, the disinvestment process was started at the 20th iteration and iterated until the minimally connected network was derived. After the disinvestment process was terminated, the network growth process continued until it became stable again. Note also that the minimal number of links to be eliminated depends on the number of axes in a symmetric network. For example, in a complete network with 3 axes, up to 8 link agents, i.e., 16 one-way links will be removed at a time.

9.5.3 Experimental results

To illustrate the dynamics of network topologies, three snapshots of test networks in each experiment are displayed, including the initial network, an interim network, and the final minimally connected network. A series of topological measures defined in Section 9.3, including the gamma index (γ), the measure of network density (D), the measure of speed entropy (H), the Gini index (G), and the measures of connection patterns are calculated for the networks. To reduce the running time, the measures were computed every five iterations. As most of the networks have rather complicated topologies, the links are classified into arterials and collectors according to their absolute speeds (links with their speeds above 10 are defined as arterials while those with speeds below 10 are defined as collectors), and the measures of connection patterns are computed only for the arterials whose topological patterns are of greater importance in a network.

9.5.3.1 Experiments on grid networks

The snapshots of an evolving grid network in asymmetric (Experiment 1) and symmetric (Experiment 2) scenarios are displayed in Figure 9.6. As can be seen, both scenarios generate hierarchical network topologies over time, but an asymmetric network evolves into a tree-like structure with cul-de-sacs, while a symmetric network evolves into four symmetric sectors connected by a ring in the center.

9.5.3.2 Experiments on complete networks

The snapshots of Experiments 3a and 4 are displayed in Figure 9.7, starting from the 4×4 complete network with random and uniform speeds respectively. Figure 9.8 displays different minimally connected networks derived from Experiment 3a, 3b, and 3c for comparison.

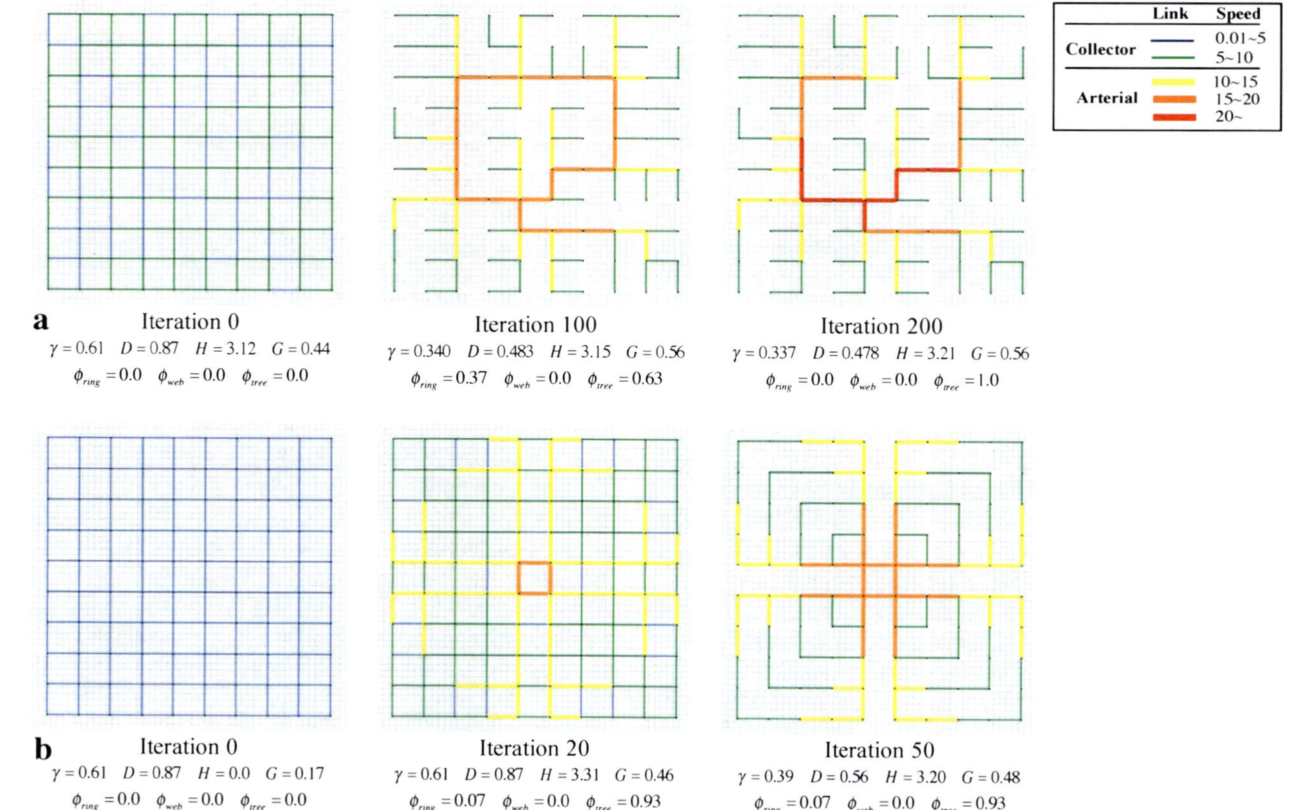

Fig. 9.6 Snapshots of grid networks in (a) Experiment 1 and (b) Experiment 2 with topological measures. According to the measures of connection patterns, typical connection patterns of arterials are identified at different stages, including rings (Iteration 100 of Experiment 1; Iterations 20 and 50 of Experiment 2), hub-and-spoke (Iteration 200 of Experiment 1), and cul-de-sacs (Iterations 150 and 200 of Experiment 1). The legend bar applies to idealized networks in other figures as well.

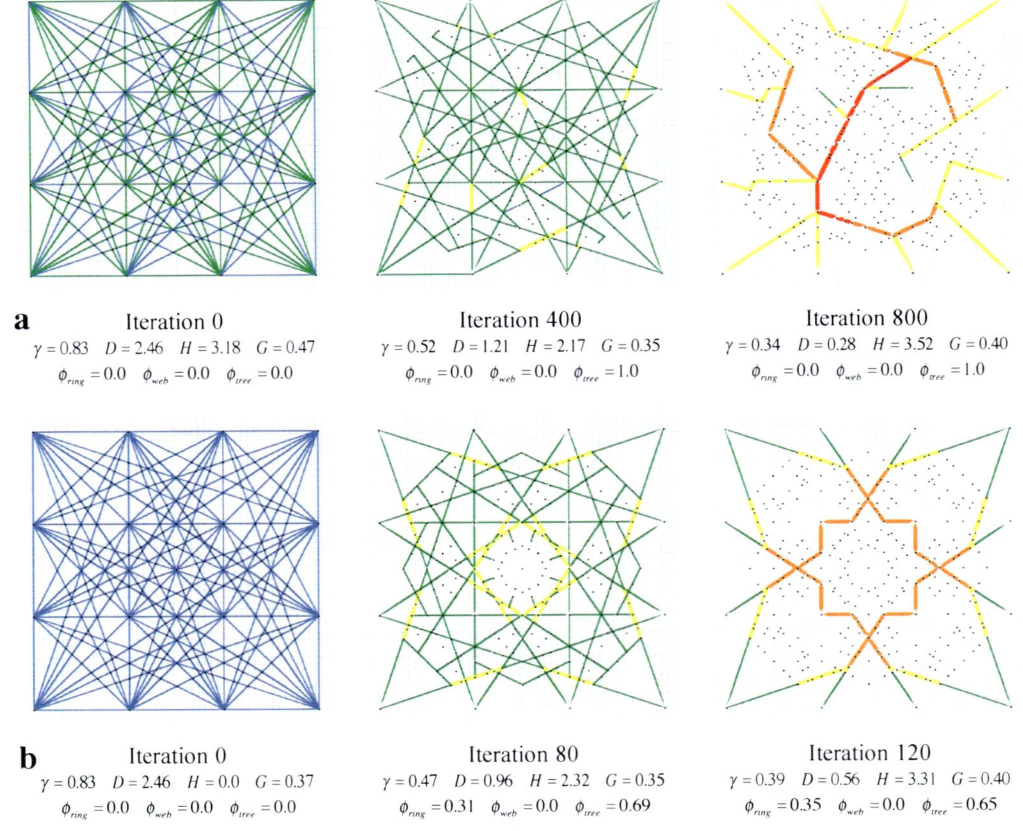

Fig. 9.7 Snapshots of complete networks in (a) Experiment 3a (Experiments 3b and 3c are identical except for different random starting speeds)and (b) Experiment 4. The asymmetric network eventually evolves into a hub-and-spoke structure with faster links connecting hubs in the center, and slower links between hubs and terminals in the suburbs; the symmetric network, on the other hand, evolves into a ring. Both experiments, however, disclose a similar dynamic process: as less used links are removed, more traffic is directed to remaining links among which the ones on more used routes collect more revenue and improve themselves over iterations into arterials, which are initially scattered in the network, and then emerge into a contiguous arterial network.

The topological change of the networks can be further corroborated by the fluctuations of topological measures over iterations. Taking Experiment 3(a) as an example, Figure 9.9 plots the measures of network topology from different dimensions. In Figure 9.9(a), the disinvestment process is characterized by a continuous decrease in connectivity (the gamma index), as the least used links are repeatedly removed from the network. Not surprisingly, the fluctuation of network density displays a similar downward-sloping pattern. The dynamics of speed entropy, as shown in Figure 9.9(b), displays more fluctuations: the initial disordered status with a random distribution of link speeds corresponds to a high initial value of speed entropy (3.18). As some links at lower levels are abandoned while others become faster and enter higher levels, the entropy slopes downwards until about the 400th iteration, when most links serve as collectors operated at a speed below 10. Then the entropy gradually increases from about 2.10 to 2.60 because more and more links are improved to arterials. When the network shrinks close to the minimal size, a substantial jump of the entropy is observed at the 780th iteration. As can be seen, the entropy measure reflects how the heterogeneity of the structure changes over time as autonomous links develop into different hierarchies. Figure 9.9(c) displays the change of the Gini index, which indicates the concentration of traffic on the netowrk. The curve slopes down as links with lowest volumes are removed and traffic becomes less concentrated across remaining links, while a substantial increase is observed at the 780th iteration. Figure 9.9(d) depicts the change of the connection patterns of arterials in terms of their relative significance since the first arterial appears at the 170th iteration. As more and more links improve themselves into arterials, individual arterial links that appear at scattered locations on different routes eventually merge into a contiguous network. The arterial network becomes more and more connected in the beginning, reflected by a decrease of treeness and an increase of webness, since the first ring appears in the 530th iteration and the first web in the 535th iteration. When all the remaining links become arterials and weaker arterial links have to be removed from the network, however, the arterial network shrinks and eventually evolves into a hub-and-spoke structure. In a reverse process, the treeness increases while the webness and ringness decrease. It is worth noting that the collapse of the last ring (when measure of ringness drops to null) occurs also at the 780th iteration, when speed heterogeneity (entropy) and traffic concentration (Gini) substantially increase. This could be explained by the fact that the significant transformation of network structure leads to the redirection of a large volume of traffic on the network when a few links remain.

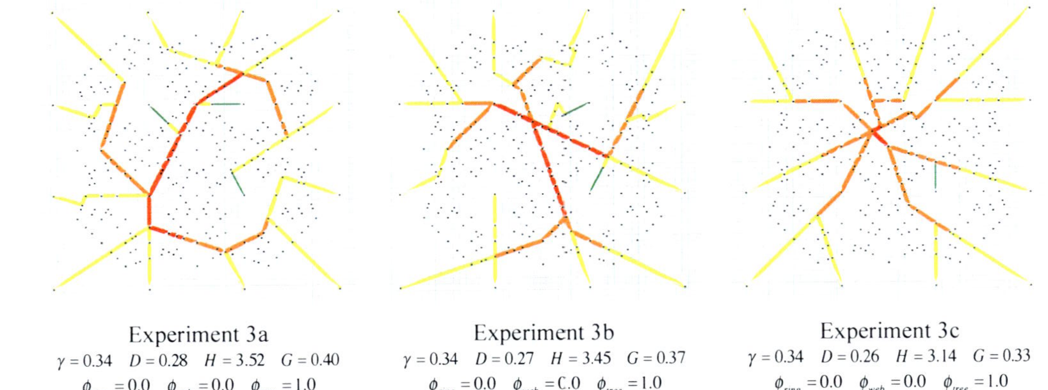

Fig. 9.8 Emergent networks of (a) Experiment 3a, (b)Experiment 3b, and (c) Experiment 3c with different sets of random initial speeds. The three experiments interestedly result in hub-and-spoke networks with quite similar measures of topological attributes.

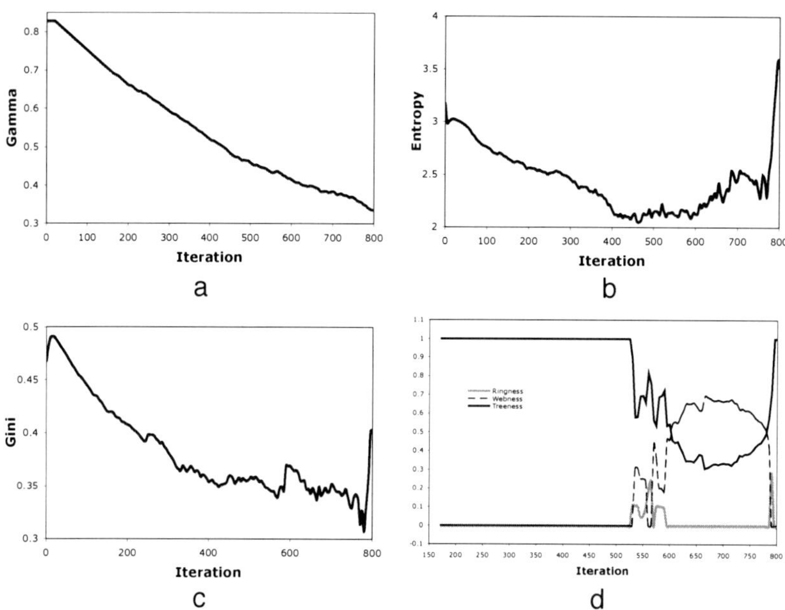

Fig. 9.9 The fluctuations of topological measures in Experiments 3a from the dimensions of (a) Connectivity; (b) Heterogeneity; (c) Concentration; and (d) Connection patterns.

9.5.3.3 Experiments on hexagon networks

The snapshots of Experiments 5 and 6 are displayed in Figure 9.10. Note that the original hexagon network is designed with nodes evenly distributed on a homogeneous landscape. Initially, different levels of nodes (differentiated by number of connections) have different scopes of "market", within which the cells are allocated to the most dominant place. Without considering the nodes on the border of the network, the primary nodes (centers of hexagons) are allocated more cells. Thus they are expected to represent more important places which serve more traffic. As the network evolves, however, the function and relative importance of places changed. As arterials connected into spokes or beltways, minor places along these links are reinforced, eventually serving a majority of though traffic. This finding suggests that instead of assuming a static landscape as in Central Place Theory, the potential impact from an evolving transportation network on the land use pattern needs to be considered for a more realistic representation of network dynamics.

Another interesting finding is the temporal change in link spacing. In the initial hexagon network there exist 12 directions for traffic from each hexagon center. Along 6 directions, parallel links have smaller spacing. When the network becomes less dense, as we can observe in Figure 9.10(a) (Iteration 500) and Figure 9.10(b) (Iteration 200), the spacing of parallel links along these directions is enlarged. This can be explained by the fact that autonomous links on two parallel routes have to compete for through traffic; since the initial hexagon network is dense (as we can see by comparing it to initial grid and complete networks with regard to connectivity and density) and the traffic is insufficient to support both routes, links on one route have to degenerate while those on the other survive. This finding suggests that link spacing in an urban transportation network is another emergent property of spontaneous network dynamics.

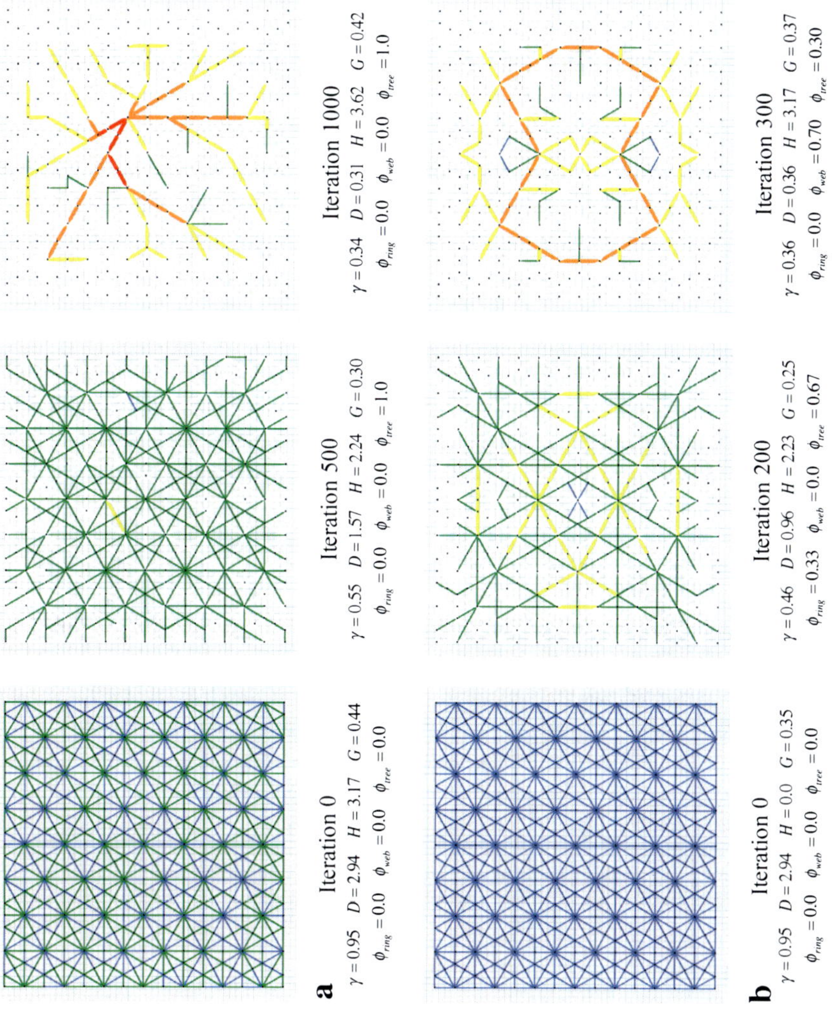

Fig. 9.10 Snapshots of hexagon networks in (a) Experiment 5 and (b) Experiment 6. Experiment 5 starts from an asymmetric hexagon network and evolves into a hub-and-spoke structure which holds topological attributes similar to those emerge in the complete network. Experiment 6 starts with uniform initial speeds. As the network grows, scattered arterials emerge and connect into a ring at the 200th iteration, and finally expand into a mesh-like network.

9.5.4 Sensitivity tests

Descriptions of model parameters and their specified values are outlined in Table 8.1. As the values of model parameters are specified, changing these values may lead to different results. Therefore, a sensitivity test is taken by repeating the experiments with different parameter values. Taking Experiment 1 as an example, the heterogeneity measures of resulting network topologies with different values of a parameter are depicted in Figure 9.11.

Increasing the specified walking speed (v_0) from 0.01 to 1.0 decreases the access cost to the network, thereby increasing the travel demand on the network. Since the "weakest-link" criteria is concerned only with the relative travel demand across links, however, changing the walking speed did not affect the topological attributes of emergent networks significantly.

A higher toll rate (τ) means more revenue collected on a link with the same amount of through traffic. Raising the toll rate from 1.0 to 10.0 significantly increases the average speed of links in equilibrium when the disinvestment starts, therefore resulting in a higher measure of entropy as compared to the default settings. It needs to be noted that the toll rate in this model is externally specified and there is no demand feedback on it. In reality we would expect revenue to decline when toll is sufficiently high, so we may not have a revenue maximizing toll within the range set above. A more sophisticated model could address this issue by endogenizing toll setting in response to the varying demand on a link. Chapter 14 will discuss this in greater details.

A lower flow coefficient (α_2) in the cost function indicates a higher economy of volume in maintenance. Decreasing the coefficient from 0.75 to 0.25 favors the links with higher volumes, which essentially intensifies the differentiation of links during the dynamics. This is corroborated by the observation of higher entropy throughout the experiment.

The speed improvement coefficient (β) specifies how intensively each individual link invests or dis-invests in the infrastructure in response to its profit or deficit. Changing the speed improvement coefficient, however, did not change the link speeds in equilibrium when the disinvestment starts. Therefore, no significant effect on the emergent networks has been observed after increasing this coefficient from 1.0 to 2.0.

9.6 Findings and concluding remarks

While fully recognizing the critical roles that regulators and planners have played in the formation of transportation networks in recent decades, this chapter, following Chapter 8, represents the evolution of surface transportation networks as a spontaneous process which is played out as the outcome of completely localized decisions made by independent infrastructure users and suppliers. The SOUND model, enabling a variable network topology in a degeneration process based on the myopic

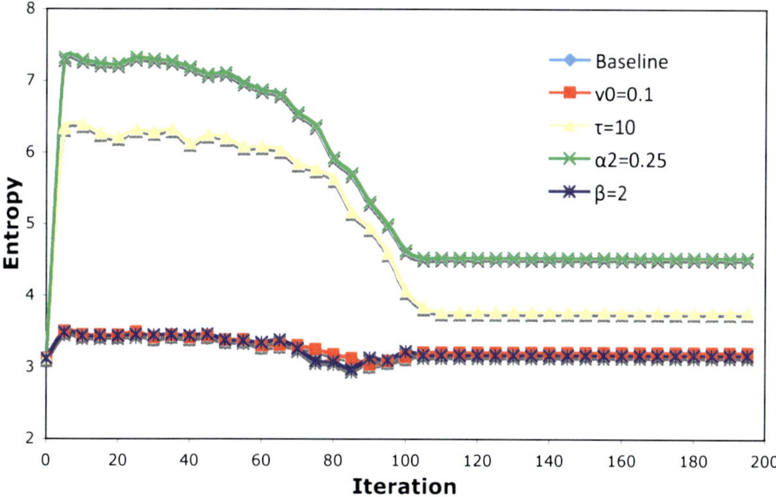

Fig. 9.11 The sensitivity of topological dynamics on model parameters with respect to heterogeneity. Each experiment changes the value of one parameter with others being equal. The experiments display different fluctuations of entropy.

"weakest-link" heuristic, provides an innovative understanding of how these decentralized decisions could be translated into facilities on the ground. Additionally, this study introduced spatial measures that helped quantify the topological features of transportation networks and trace their structural changes over time.

The model was validated using historical data extracted from the Indiana interurban network during its decline phase. Statistical analyses disclosed that the model performed well in reproducing the sequence of link abandonment in the interurban network, as well as the temporal change of its topological attributes.

The model was then applied in simulation experiments to different idealized network structures with different initial conditions. Typical connection patterns such as rings, webs, hub-and-spokes, and cul-de-sacs were observed in the evolving networks. The observation that the same type of connection patterns may arise from different initial conditions, or from different network structures, even based on completely decentralized decisions, suggests that surface transportation networks possess robust topological properties that spontaneously emerge from the interaction of demand and supply. Moreover, the heterogeneity measures such as speed entropy and the Gini index revealed the spontaneous organization of network hierarchies in a variable network, which agrees with the finding from the preceding chapter with fixed network topologies.

The spontaneous change in network spacing between parallel links provides further evidence for the interaction between demand and supply during network evolution. The enlargement of spacing is accompanied with the abandonment of links on alternate routes, which is a natural reaction to the over-competition for traffic in a dense network. This finding also indicates a future direction of model enhancement to explicitly include cooperation and competition between decentralized infrastructure suppliers in different ownership structures.

The rise-and-fall of places in terms of their relative importance is also observed as the transportation networks evolve, suggesting the impact of network evolution on land use. Following an abundant literature on integrated transport-land use analysis, we will examine the formation of places coupled with network diffusion in Chapter 11, and model the co-evolution of transportation and land use in Chapter 12.

Above all, this research has demonstrated that, with both empirical and simulation evidence, a sequential degeneration of an ultra-connected transportation network driven by localized supply decisions that did not follow an optimal design could reproduce the basic topological features of surface transportation networks. The findings provide clear evidence that the formation of surface transportation networks exhibits spontaneous organization.

Chapter 10
Sequence

10.1 Introduction

As of 2010, Hennes & Mauritz AB (operating as H&M), a Swedish clothing company, has around 370 stores in Germany, accounting for approximately a third of its global sales. Figure 10.1(a) visualizes the geographical locations of German cities in green and H&M stores in red. Known for the fast fashion clothing offerings, the H&M salesman is facing the immense task to visit each store exactly once in a shortest possible tour given the pairwise travel costs between one and another.

This is one of the Traveling Salesman Problems (TSP) studied in operations research and theoretical computer science. For its computational difficulty in instances with hundreds to thousands of cities, a large number of heuristics methods are known. Among them is the nearest neighbor algorithm, or so-called greedy algo-

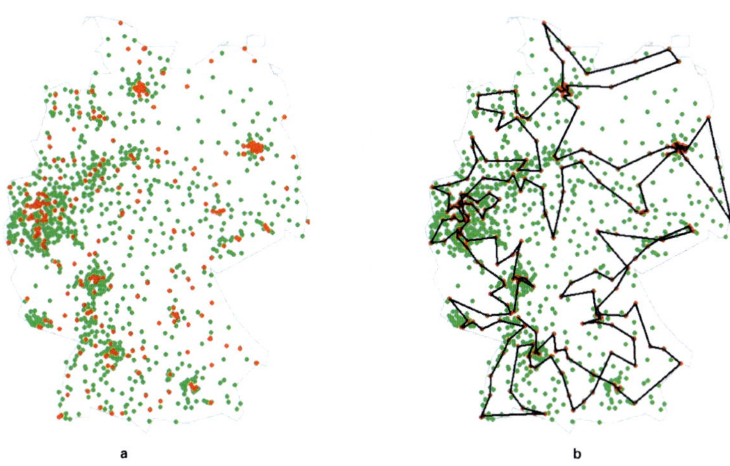

Fig. 10.1 H&M Travel Salesman Problem. Source: Eugster (2010)

rithm, which quickly yields an effectively short route by letting the salesman choose
the nearest unvisited city as his next move. Figure 10.1(b) displays a solution to the
H&M traveling salesman problem using the greedy algorithm. Conducting experiments with randomly generated instances of nine classes (each with 24 instances),
Gutin and Karapetyan (2008) demonstrated that the greedy algorithm on average
yields a path 20.8% longer than the best known solutions.

The greedy algorithm follows the problem solving meta- heuristic of making the
locally optimal choice at each stage with the hope of finding the global optimum.
Chapter 9, in the efforts of validating the SOUND model, disclosed that a simulation based on the intuitive "weakest-link" heuristic performs well in predicting the
sequence of link abandonment during the decline phase of the interurban network
in Indiana. This finding supports the notation that transportation development could
be described as a discrete process by which changes to infrastructure networks are
realized based on myopic, local optimal decisions.

Extending this idea, this chapter investigates whether and to what extent the geographical expansion of a transportation network can be represented as a sequence
of link additions to the network based on myopic, local optimal decisions made in
discrete time periods. This chapter will proceed with a graph-theoretic definition
of "link addition problems". A model of incremental network connections is then
outlined, followed by its application to simulating the expansion of the skyway network in downtown Minneapolis. The concluding remarks highlight our findings and
indicate the implications.

10.2 Model

10.2.1 Incremental connection problem

The sequential deployment of a surface transportation network over space and time,
which assumes the form of "link addition problems" that have been previously defined in transportation geography (Haggett and Chorley, 1969), deals with how links
will be added among a set of fixed nodes to create an efficient network. Suppose
there is a complete graph $G = \{V, L\}$ that comprises a finite set of potential vertices
V and potential edges E, and a set of established places pre-specified as $P \subseteq V$, each
occupying one and only one vertex in the graph.

Established places could be connected by transportation links (e.g., roads) in
one continuous network or in separate subnetworks ($G_m = \{V_m, L_m\}, m = 0, 1, 2, ...$),
supposing an established transportation link is represented by one or a series of
consecutive directed edges. A subnetwork holds the following three properties:

- A subnetwork must be a subset of the compete graph G:

$$V_m \subseteq V, L_m \subseteq L \tag{10.1}$$

- A pair of subnetworks shares no vertices or edge:

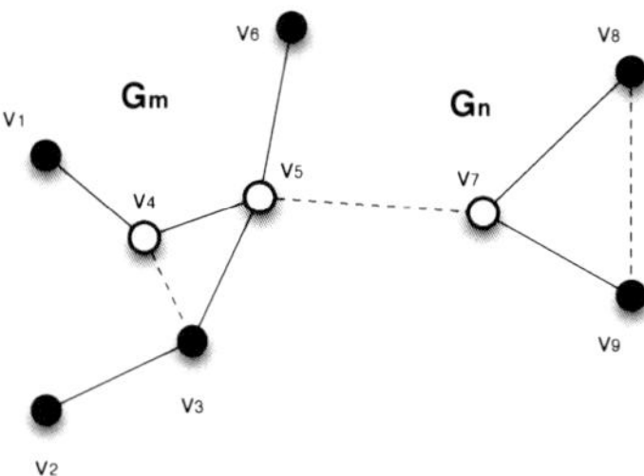

Fig. 10.2 A graphic example of external versus internal connections

$$V_m \cap V_n = \emptyset \tag{10.2}$$

- A subnetwork contains at least one established place (an isolated place without any connection can be viewed as a scalar subnetwork):

$$V_m \cap P \neq \emptyset \tag{10.3}$$

Additions of transportation links between places can be either internal or external connections. An internal connection is defined as the connection made between two vertices that belong to the same subgraph. On the other hand, if a connection is made between two vertices that belong to different subgraphs, it is referred to as an external connection. An internal or external connection represents a series of nodes and two-way edges that consecutively connect along the geographical shortest-distance path in the complete graph. For illustration, Figure 10.2 presents a graphic example where solid lines represent established links, dashed lines proposed connections, dark dots established places, and white dots intersections. As can be seen in the figure, connections $v_3 - v_4$, $v_8 - v_9$ are internal connections while $v_5 - v_7$ is an external connection. In essence, an external connection adds a branch link in the network while an internal links adds a circuit link.

Based on the above specification, an incremental connection process is defined below:

- Step 0: Start with the complete graph G and a set of unconnected places that belong to separate subgraphs.
- Step 1: One connection is made at a time.
- Step 2: Two separate subgraphs merge when an external connection is made connecting them.
- Step 3: As the process goes on, a connected network of places and established links may eventually emerge. Steps 1 and 2 are repeated until the topology of the established network remains unchanged based on pre-specified stopping criteria.

10.2.2 Model framework

In this section, System Of Network Incremental Connections (SONIC) is constructed to implement the above defined incremental connection process by which a surface transportation network is deployed on a link-by-link basis between a given set of established places. The interconnection of SONIC's component models is illustrated in the flowchart of Figure 14.2.

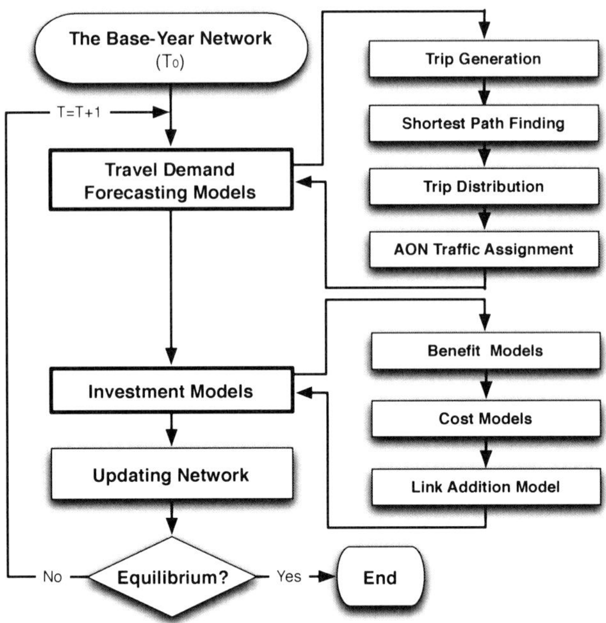

Fig. 10.3 Model framework of SONIC: System Of Network Incremental Connection

Similar to the SONG model and the SOUND model presented in previous chapters, travel demand models predict aggregate traffic across a given network by following the traditional planning steps of trip generation, trip distribution, and traffic assignment. Modal split is skipped by assuming a single mode of trips.

Different from the previous models that pre-specify number of trips originating in and destined for each land use block and assign these trips to the nearest nodes, the SONIC model assumes a set of established places, and estimates travel needs by associating trip generation and trip attraction with land use activities pre-determined for each place. For simplicity, a linear relationship is assumed between trips and activities as follows:

$$O_i = \xi_0 + \sum_k \xi_k P_i^k \tag{10.4}$$

$$D_i = \psi_0 + \sum_k \psi_k P_i^k \tag{10.5}$$

Where O_i and D_i represent trips originating in and destined for Zone i, and P_i^k denotes the intensity of a particular type of land use activity in Zone i.

A doubly constrained trip distribution model predicts the trip ends between each pair of origin and destination places. Similar to the formulation in Equation 8.3, trips going from place i to place j is estimated as:

$$N_{ij} = K_i K_j O_i D_j f(t_{ij}) \tag{10.6}$$

where K_i and K_j represent K- factors to be estimated in the model. The generalized travel cost t_{ij} between origin place i to destination j is calculated along the least-cost route as:

$$t_{ij} = \begin{cases} \sum_a \delta_a^{ij}(t_a + \tau_a/\upsilon) + 2t_0 & i \neq j \\ t_0 & i = j \end{cases} \tag{10.7}$$

where δ_a^{ij} indicates if link a is on the least-cost route. Variable t_a represents the time cost for traversing this link, while τ_a the monetary cost (if any). Variable υ denotes the average value of time that converts the monetary cost into equivalent time cost. Variable t_0 denotes a universal intra-zonal travel time.

As in the previous chapters, congestion effects are not considered in this analysis. No capacity constraint is assumed on individual links, and thus an all-or-nothing (AON) traffic assignment is employed, by which all the trips between two places are assigned onto the least-cost path.

Investment models evaluate the benefits and costs associated with each potential addition to an existing network, and predict the strategic investment decisions made by infrastructure provider(s). Unless in a highly speculative environment like what we have seen in the cases of Indiana interurbans and Twin Cities streetcars, a rational provider considers both benefits and costs associated with each investment choice. That being said, it is intuitive to posit that the most cost-effective connection will be added to the network first, which, in contrast with the "weakest-link" heuristic studied in the previous chapter, is referred to as the "strongest-link" assumption in this analysis. The "strongest-link" heuristic expands a network on a link-by-link basis: for each time period, an exhaustive set of potential link additions is identified. Among them the one that would realize the highest benefit-cost ratio is built.[1] After a round of investments are made, the network is updated, the time period is

---

[1] For simplicity, costs are estimated by assuming that the general cost associated with a link addition (including costs of right-of-way acquisition, construction, maintenance, among others) is proportional to the length of the link, and that the unit cost is uniform across the network. The benefits of a link addition, on the other hand, need to be defined and evaluated on a case-by-case basis, as the empirical findings from Chapters 4 and 5 suggested that the beneficial effects of a potential link addition differ by the ownership structure (public versus private, centralized versus decentralized) of its provider(s).

incremented, and the whole process is repeated until candidates for potential link additions are exhausted.

10.3 Simulation

Chapter 4 described the incremental expansion of the skyway network between blocks of downtown Minneapolis. The empirical analysis revealed that accessibility is a significant predictor of link additions to the skyway network. The connect-choice logit model further suggested that despite all the specific physical, economic, and regulatory factors that affected the investment choices, in the long run property owners seemed to follow some local, myopic optimal rules in the attempt to build a skyway that maximizes the accessibility to immediately connected blocks.

Based on these empirical findings, this section develops a simulation which applies the SONIC model to "predicting" the incremental growth of the skyway network, and examines to what extent the predicted sequence of link additions matches with the observed. The model expands the skyway network based on the "strongest-link" heuristic, by which one and only one skyway link is constructed per turn, and this link is the strongest one among all the candidates in terms of providing the highest increase in accessibility for the two blocks it connects.[2] Built upon this simple assumption, the model obviously sacrifices the considerations of more context-specific factors that may affect the connect-choice in the skyway system. The purpose of this model, though, is to examine to what extent simulation based on the myopic local optimality heuristic associated with accessibility can reproduce the observed sequence of network expansion. Specifically, this model includes the following four steps, which are similar to those for the connect-choice model in Chapter 4 except for the last step:

- Step 0: Find all the potential link additions for network expansion.
- Step 1: Calculate the least travel time paths from one block to every other block provided that one candiate link is hypothetically improved from a non-skyway link to a skyway one.
- Step 2: Evaluate the changes in accessibility to immediately connected blocks provided that one candidate link is hypothetically built into a skyway link. Repeat Step 1 and Step 2 for each candidate link.
- Step 3: Implement the construction of the "strongest" candidate which provides the highest possible increase in accessibility for the two blocks it immediately connects (as opposed to the actually built candidate(s) as in the connect-choice model). The improved travel speed (and thus travel time) on this link is accordingly updated.

---

[2] With approximately equal block size across downtown Minneapolis, lengths of skyways connecting these blocks are about the same. For simplicity it is assumed that the cost associated with each skyway addition is equal. Regarding increases in accessibility as the benefit of skyway additions, the "strongest-link" addition is indeed the most cost-effective one.

The following assumptions are made in implementing the SONIC model for the skyway case:

- One and only one two-way skyway link is built for each iteration.
- The link that is built is the "strongest link" among eligible candidates in terms of providing the highest additional accessibility for the two immediately connected blocks.
- The eligible candidate links include all the links that have the potential to be built as skyway links, but have not yet been built by the year of examination. Candidate links cannot connect to any restricted blocks.
- The simulation starts with one already established link, which was the first skyway built in 1962 connecting the Northstar Center with the Northwestern National Bank. The simulation ends when the candidate pool is exhausted.
- Consistent with the analysis in Chapter 4, the intensity of land use activities in a block is proxied by the total area of buildings located in this block. Parameters in Equations 10.4 and 10.5 thus correspond to trip production and attraction rates that can be obtained from Institute of Transportation Engineers (1997).
- Since the area of buildings changed over years, built area of each block in simulation is estimated by interpolation of actual data from different years.[3]
- Consistent with Chapter 4, the gravity function in the trip distribution model is defined in Equation 4.2 in a quadratic form.

10.4 Results

Based on the "strongest-link" assumption, the simulation generated a sequence of skyway additions from scratch, which was then compared to the observed chronological sequence of skyway connections retrieved from historical data. A Spearman's rank-order correlation test, which assesses how well an arbitrary monotonic function describes the relationship between two variables, was performed to examine how well the simulated and observed sequences of skyway connections are correlated to each other, without making any assumptions about the frequency distribution of the two sequences (Higgins, 2003). The results of the statistical test show that there is a significantly strong correlation (0.591), demonstrating that the SONIC model based on the "strongest-link" heuristic can replicate the observed sequence of skyway link additions to a significant extent.

---

[3] The skyway analysis in Chapter 4 revealed that the total length of constructed skyways (which can be approximately by number of established skyways assuming equal skyway lengths) is highly correlated with year. Thus the interpolation was implemented in simulation using number of established skyway links. For example, if there are 25 established skyway links at the beginning of a simulation time period, and it is known there were 22 established skyways in 1983 and 28 in 1984, then the square footage of this block in the specific simulation period is estimated by averaging the the actual square footages in 1983 and in 1984 using the weights calculated from numbers of established links.

Table 10.1 Spearman correlation coefficients calculated under alternative hypotheses

	Spearman Coefficient	P-Value
Strongest-link hypothesis	0.591	0.00
Random-link hypothesis		
Random Seed 1	-0.166	0.19
Random Seed 2	-0.136	0.28
Random Seed 3	-0.205	0.10
Random Seed 4	-0.008	0.97
Random Seed 5	-0.161	0.20

In order to examine the robustness of the Spearman rank-order correlation coefficient obtained based on the "strongest-link" hypothesis, an alternative hypothesis was tested by which the next skyway connection is instead randomly drawn from the candidate set. We ran the simulation based on the alternative hypothesis and reran the Spearman correlation test correlating the sequence of random link additions to the observed sequence. The process was repeated five times with different random seeds, and the results are summarized in Table 10.1. As can be seen, none of the Spearman coefficients resulted from the random link-addition process are statistically significant, indicating that a random sequence of link-additions is independent from the observed sequence. Using random-connect results as a benchmark, the Spearman coefficient calculated based on the strongest-link hypothesis not only shows the right sign, but is statistically significant. This experiment provides further evidence that accessibility is a significant predictor of link-connect choice in the skyway network, and that simulation based on a myopic "strongest-link" hypothesis performs well in predicting the sequence of link additions in the network.

The performance of the simulation model can also be evaluated by comparing the topological features of the resulting network in simulation to those observed in reality. For this purpose, a series of topological measures were computed to evaluate the structural features of the skyway network throughout the course of network expansion, including the gamma index (γ) and three measures of connection patterns (ϕ_{ring}, ϕ_{web}, and ϕ_{tree}), which have been introduced in the preceding chapter.

To this point, the simulation of sequential connect-choice decisions in the skyway system has been conducted with a set of candidate links that includes all the links that are eligible for skyway construction, whether or not a skyway was actually established. For the purpose of comparison, the simulation was rerun with an alternative set of candidate skyway connections comprised of only links that were actually established as skyways by the year of 2002, all else being equal. Selected topological measures were calculated for the actual network extracted based on historical observations over years until 2002, and for the simulated network based on the "strongest-link" heuristic at the end of each simulation period. Measures in observation versus in simulation were then compared as shown in Figure 10.4 when the same number of skyways were added to the network.

It came as no surprise that measures of the same topological feature in observation and in simulation converged in the end, as the simulated network eventually evolved into the 2002 network. It is interesting, though, to observe the fluctuation of a topological measure in simulation in comparison to that in observation throughout the process of network expansion. The gamma index in both cases demonstrates a general trend of gradual increase, indicating the increased interconnection as more skyway connections were included in the network. The measure of ringness in simulation indicates a ring with significant size (ringness equals 0.28) emerged when 28 skyway connections were added to the network. This is in agreement with what happened in reality, a ring of prominent magnitude (ringness equals 0.44) emerging at a slightly earlier point of time. Though the simulation did not predict a small ring that emerged later in reality. The measures of webness and treeness in both cases follow similar zig-zag patterns over time. When a branch skyway link is added connecting an isolated building to the network, the webness of the network decreases while the treeness increases; when a circuit link (either a ring link or a web link) is added between two already connected buildings, the changes to webness and treeness occur in the opposite direction. In general, the webness from simulation is larger than that in observation, suggesting the simulation model developed a skyway network that is more circuitous than the real one. This can be also observed with the measure of treeness. It should be noted that in this study data points are not sufficient to perform the Two-sample Wilcoxon rank-sum (Mann-Whitney) test which we used in the last chapter to examine the significance of topological similarities between the simulated and the observed Interurban network in Indiana. However, the comparison of topological measures still demonstrates that simulation based on a simple heuristic associated with accessibility can approximate the topological transformation of the real skyway network in Minneapolis.

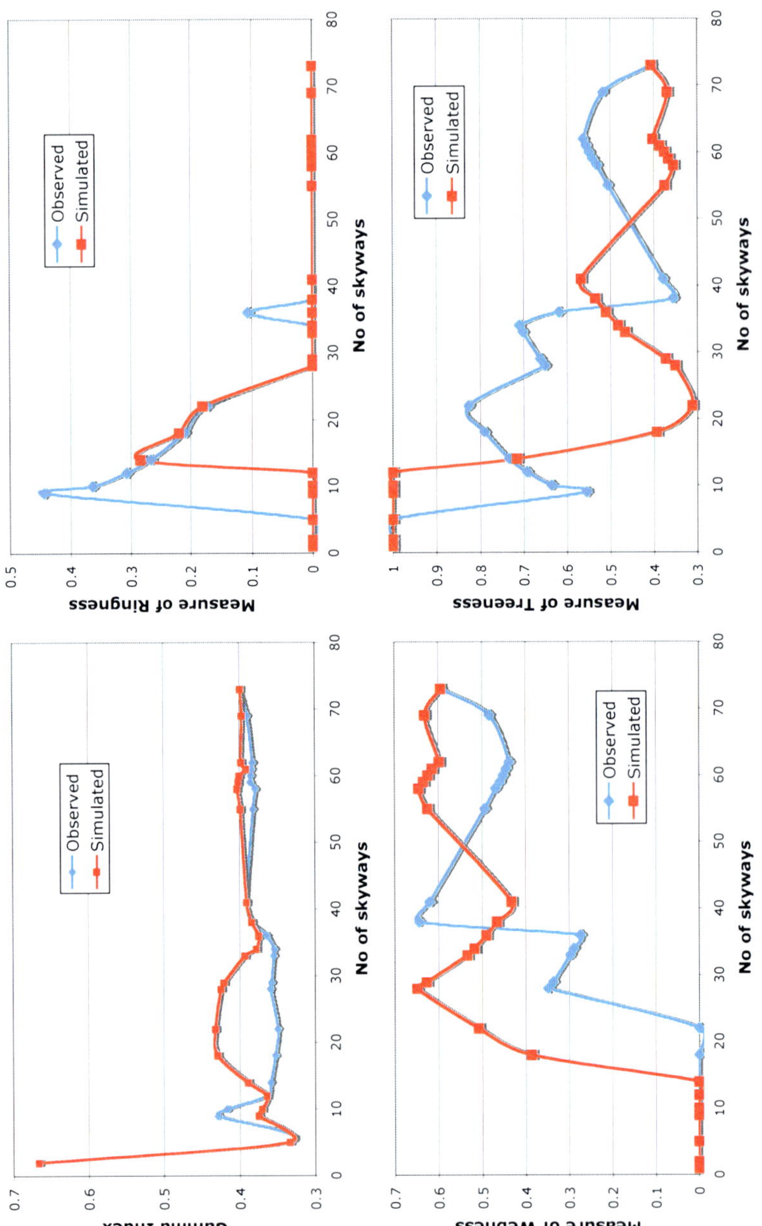

Fig. 10.4 Topological measures of the Minneapolis skyway network in observation *vs.* in simulation

10.5 Findings and concluding remarks

This chapter, inspired by the greedy algorithm, represents the geographical expansion of a transportation network as a sequence of link additions to the network based on myopic, locally optimal decisions made in discrete time periods. A System Of Network Incremental Connections (SONIC) was constructed to create a network link by link based on the "strongest-link" heuristic, by which one and only one link is added to the network per turn, and this link is the most cost-effective one among all the potential candidates. Based on the empirical finding from Chapter 4, and assuming the cost of link additions is identical, the model was applied to the skyway network in downtown Minneapolis adding the strongest link per turn in terms of providing the highest increase in accessibility for the two blocks it immediately connects. Statistical analyses showed that this model "predicted" a sequence of link additions to the skyway network that replicates the observed sequence to a significant extent. Comparison of topological measures in simulation versus in observation further demonstrated that simulation based on a simple heuristic associated with accessibility can to some extent reproduce the topological features of the real skyway network.

This research simplifies the decision-making of transportation investment in several respects. On the demand side, congestion and additional monetary costs for travelers (such as pricing) on a network are not considered. On the supply side, the model only allows construction of new links but not maintenance/expansion of existing links, while an intelligent provider should have considered both maintenance and new construction, or even been able to rank the priorities against each other. In addition, the provider(s) in this model can only optimize (though based on local, myopic optimal rules) the location of next link addition, but not its service level (such as capacity or design speed) subject to budgetary constraints. More importantly, this research only examined link addition decisions in the specific skyway case under local, private provision, while investment decisions may be different under different economic or regulatory regimes, as disclosed in Chapters 4 and 5.

Despite these limitations, this research provides an important insight into a discrete representation of network growth, and discloses that even based on local, myopic optimal decision-making, an incremental connection process can reproduce the observed sequence of link additions in a network to a significantly extent. Extensions to the SONIC model will find important applications in the subsequent chapters of this book. Chapter 11 will integrate the incremental network connection process with place formation, and use the integrated model to re-examine the First Mover Advantage (FMA) question posed in Chapter 7 in a simulation environment. Adopting the basic framework of the SONIC model, Chapter 15 will introduce a more realistic representation of the incremental connection process, and examine the development of transportation networks under alternative governance structures.

Part IV
LAND USE

This part investigates the development of transportation networks while endogenizing the evolution of surrounding land use simultaneously.

Chapter 11
Network Diffusion and Place Formation

11.1 Introduction

Extending the discussions in Chapter 7, this chapter revisits the topic of first mover advantages. Chapter 7 examined the existence or absence of first mover advantages in the transportation sector based on empirical observations. Examining the empirical case of London rails suggested the existence of first mover advantages in surface transportation networks, and revealed its close relationship with the spatio-temporal location of stations and network connectivity. Analysis of the global aviation system indicated a significant persistence of airlines at hubs, suggesting accumulative first mover advantages for early established airports. The case of Twin Cities roads, on the other hand, suggested that an accurate treatment of first mover advantages requires the integrated modeling of transportation and land use in an evolutionary context, for which the empirical approach is largely limited due to the scarcity of extensive *ex post* data.

In recent decades, the theoretic research on first mover advantage has gained momentum since its application in the fields of industrial marketing and management. Since the publication of the seminal paper by Lieberman and Montgomery (1988), who defines first mover advantages in terms of "the ability of pioneering firms to earn positive economic profits" in a competitive market, a broad literature has been dedicated to exploring the mechanisms that confer advantages on first mover firms in specific market segments (Kerin et al., 1996; Makadok, 1998; Rahman and Bhattacharyya, 2003; Mittal and Swami, 2004). Extending Lieberman and Montgomery (1988), Mueller (1997) further identified a number of sources for first mover advantages in industrial markets, which he considered both demand and supply related. Demand related inertial advantages include set-up and switching costs, network externalities (effects that one user of a good or service has on the value of that product to other people), and buyer inertia (a consumer's resistance to change buying choices) due to habit formation or uncertainty over quality, while supply-related efficiency advantages include set-up and sunk costs, network externalities and economies, scale economies, and learning-by-doing cost reductions. Controver-

sially, disadvantages may also be involved for first movers. For example, early entrants to a market may miss the best opportunities and acquire the wrong resources,
obscured by technological and market uncertainties during the early stage of the
market (Lieberman and Montgomery, 1998).

Transportation systems, however, exhibit many properties that are distinctly different from industrial markets, and the first mover theories discussed above may not
apply. For instance, in contrast to firms in a free market that are private and highly
competitive in nature, transportation infrastructure is largely public, so providers of
transportation infrastructure may be neither profit-pursuing nor competing with each
other.[1] That being said, the notion of first mover advantage needs to be carefully re-
defined for the purpose of this study based on an in-depth look into the distinctive
characteristics of surface transportation systems.

This study examines first mover advantages from an evolutionary perspective,
asking if early arrival is the cause of higher connectivities and confers advantages
to transportation facilities, and if the advantages remain or change after passage of
time. Dealing with transportation systems subject to both spatial and network constraints, and shaped by interdependent economic and political initiatives in deployment decision-making processes, we examine these questions using an approach
different from the empirical analyses presented in Chapter 7, or the game-theoretical
methods adopted in the industrial economics literature. Rather, we construct an *ex
ante* model of network diffusion coupled with place formation, and analyze first
mover advantages in a controlled environment. The next section introduces the
model and proposes measures of first mover advantages, which is followed by simulation experiments and a discussion of results. The conclusion section highlights
our findings and indicates future directions of this research.

11.2 Model

11.2.1 Major assumptions

In order to examine the question of first mover advantage more rigorously, we pro-
pose an *ex ante* model of network diffusion coupled with place formation by which
first mover advantage can be defined and assessed in a controlled spatial environment. The model aims to capture the essence of locational and temporal first mover
advantages in a spatial network. To do so, we sacrifice some important, but from
our point-of-view non-critical, considerations such as land use development and
congestion.

An important simplification of our model is to treat the distribution of land use activities as exogenous and fixed through time. While fixing land uses, this model pre-

---

[1] While some transportation systems such as streetcars and interurbans (Marlette, 1959; Diers and
Isaacs, 2006) were privately developed, most current highway and transit systems were provided
with extensive public funds.

dicts the chronological formation of new places based on the relative attractiveness (proxied by accessibility) of potential locations to land use activities. As a network develops, the distribution pattern of accessibility varies accordingly, thereby affecting the formation of new places, and driving a new round of network development. By doing so the model partially captures the impact of a transportation network on urban growth in a mutual process of network diffusion and place formation. This will be further discussed in the place-formation and link-formation sub-models.

Another important simplification deals with link resizing. As a transportation network expands, its links (such as roadways and transit lines) or nodes (such as seaports and rail stations) may be resized (in most occasions with increased capacity) to accommodate the varying travel demand. In reality, resizing decisions on individual links are made in a complicated investment process that may involve different economic or political initiatives and be limited by the availability of resources and information. To simplify, this study assumes existing links being automatically resized as a network evolves to ensure free-flow travel throughout the network. This assumption is not unreasonable considering our analysis is limited to the early deployment phase of a transportation technology[2] when the issues of congestion and funding deficiency are not as significant as in its maturity stage. With this assumption, the model eliminates congestion from our analysis which may set in counteracting locational advantages of some heavily used links.

The third simplification assumes the advantage of a link / node in a transportation network is proxied by the volume of traffic flow traversing this link / node. A link / node with a larger volume of through traffic represents a more important network element in terms of serving travel needs, improving network connectivity, and promoting surrounding real estate values. Moreover, given the resizing assumption posited above that an existing link is always resized to meet the demand of free flow traffic, a link that carries a larger amount of traffic is in an advantageous position that acquires a higher level of infrastructure investment. One may argue that air pollution, visual blight, runoff and other concerns are serious nuisances associated with traffic flow, but again, limiting the analysis in the early deployment stage of a network when congestion is not as significant as in the maturity stage, this study regards spatially differentiated travel demand as a vital indicator of the locational advantage a facility gains in the network.

With the assumptions outlined above, this study represents the co-deployment of a surface transportation network and places as a bi-level iterative model. The outer loop implements a place formation model predicting locations where places form from a primitive landscape. The inner loop includes a link formation model which, adopting the basic form of System Of Incremental Connections (SONIC), deploys transportation links to incrementally connect established places. The model is thus referred to as System Of Network Incremental Connection and Place Formation (SONIC/PF).

---

[2] The deployment phase of a transportation network is defined as the period when infrastructure is deployed to connect isolated locations when the network expands over space. It corresponds to the birth and growth stages of the life cycle described in the S-curve theories (Nakicenovic, 1998; Garrison and Levinson, 2006).

The place / link formation model enables incremental place / link addition in an iterative process by which one and only one place is added in an outer-loop round and one and only one link is deployed in an inner round. The process is terminated once candidates are exhausted[3] and the network remains unchanged. The model is illustrated by a flowchart shown in Figure 11.1, and its component models explained in turn as follows.

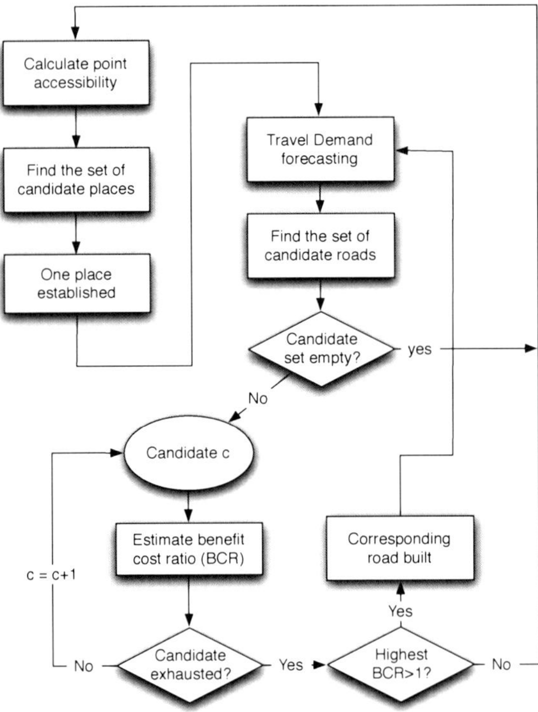

Fig. 11.1 Model framework of SONIC/PF: System Of Network Incremental Connection and Place Formation

11.2.2 Place formation model

A place formation model predicts the emergence of new places over a hypothetical space which is compromised of a nest of cells with a pre-specified distribution of

---

[3] Candidates for a potential place are exhausted when no eligible local peak cells are available; candidates for a potential transportation link are exhausted when no potential route could be deployed with a benefit-cost ratio above one. More details will be discussed later.

land use activities. It is reasonable to posit that a place first emerges where desired activities are most accessible, thus we define the attractiveness of a particular location (proxied by the centroid of a land cell) in terms of its accessibility to land use activities distributed over space. As in previous studies, accessibility is defined as the ease of reaching desired land use activities impeded by the cost of transportation. This study assumes only two types of land use activities over space: labor (workers) and employment (jobs), both located at the centroids of the cells, and defines two accessibility measures below: accessibility for a worker to reach jobs across the region ($A_{i,J}$), and accessibility for an employer to reach the labor force ($A_{i,W}$).

$$A_{i,J} = \ln\left(w_i \sum_j u_j e^{-\theta t_{ij}}\right) \tag{11.1}$$

$$A_{i,W} = \ln\left(u_i \sum_j w_j e^{-\theta t_{ij}}\right) \tag{11.2}$$

where w_i and u_i denote number of workers and number of jobs in cell i, θ is the friction factor in the gravity model, and t_{ij} indicates the generalized travel time from cell i to cell j.

Assuming one unit of accessibility to workers compensates μ units of accessibility to jobs , a composite measure of accessibility takes the following form:

$$A_i = \mu A_{i,W} + A_{i,J} \tag{11.3}$$

We posit that the likelihood of a place getting established at a particular location depends on the relative desirability (proxied by the composite accessibility defined above) of this location among all potential locations.[4] We identify "local-peak" cells as potential locations of place formation. A cell is labeled as a "local peak" when its composite accessibility is greater than all its neighbor cells on the grid. A cell is prohibited from becoming a local peak if it is located on the fringe of a region[5] or if any of its neighbor cells has already been established as a place. Local peak cells make up the choice set, among which one and only one will be established as a new place per turn. The possibility of a local-peak cell getting established as a place is determined in a logit model as follows:

$$p_{c_0} = \frac{e^{\eta A_{c_0}}}{\sum_c e^{\eta A_c}} \tag{11.4}$$

---

[4] While businesses want to be accessible both to other businesses and to residents (who are their suppliers of labor and customers) people may want to live near jobs but far from other people to maximize available space and to avoid potential competitors for jobs. That said, the assumption of representing the desirability of a potential place using one composite measure of accessibility in this analysis simplifies the matter without differentiating the locational preferences of businesses and residents. The next chapter will introduce a set of accessibility measures that are more representative of the different locational choices of businesses and residents.

[5] A cell on the fringe of a region would more likely become a local peak as it has fewer neighbors. To avoid such a bias, the place formation model eliminates any cell on the fringe from the set of potential locations.

where A_c indicates the composite accessibility of candidate c while η is the scaling factor indicating how likely a cell with greater composite accessibility gets established as a place.

11.2.3 Link formation model

A link formation model deploys transportation links to connect a given set of established places. The model replicates the SONIC model illustrated in Figure 14.2, which includes travel demand models and investment models in an iterative process.

Travel demand models predict traffic flows across an established network, which are central not only in modeling link formation as discussed later, but also in assessing the locational advantage of a link / node relative to its counterparts in the network. Consistent with preceding analyses, the model includes trip generation, trip distribution, and traffic assignment, while skipping mode choice by assuming a single mode of travel. Trip generation models are made very simple: a worker generates and a job attracts one round trip per day; a doubly-constrained trip distribution model predicts cell-to-cell trips with the decay factor set equal to the friction factor of the accessibility models presented above; since no congestion is involved in our model, all-or-nothing traffic assignment is adopted to assign cell-to-cell trips onto the lowest travel time paths between origins and destinations.

The investment models build one route (a consecutive series of transportation links between two established places) per turn in an incremental process of link additions. Assuming a central, public authority who provides transportation infrastructure to maximize both residents' accessibility to jobs and business's accessibility to workforce, the benefit of adding a route is estimated as the potential increase in the composite accessibility. The cost of deploying the route, on the other hand, is assumed proportional to the route length at a fixed distance cost rate. Extending Chapter 10, the model selects the most cost-effective route to build among all candidates. Theoretically there may exist a multiplicity of potential routes between two places. Among them the path that minimizes travel time and the path that minimizes the map distance (regardless of link speeds) represent two logical options for route choice, since the former, usually maximizing the use of existing links, requires less construction while the latter, ignoring the established infrastructure, minimizes travel distance but requires more construction. To reduce the size of the candidate set, the investment models consider only these two routes for each pair of established places.

11.3 Simulation experiments

The model is tested on a planar, otherwise undifferentiated space (except as noted below) with neither established places nor transportation infrastructure. Locations of land use activities (centroids of land use cells) are connected by primitive trails at a speed of S_l kilometer per hour. In hypothetical scenarios as shown in Figure 11.2 or Figure 11.3, centroids of land use cells are distributed on a delta grid with the same distance of D kilometers between any pair of neighbor centroids. Each centroid is the center of a hexagonal land use cell, which holds specified numbers of jobs and workers, both assumed fixed over time. Christaller (1933) and King (1985) demonstrated in Central Place Theory that activities are distributed at nodes of different levels in the hexagonal network, which represent centers of nested hexagons. In this case, centroids distributed at a distance of D (in kilometers) belong to the lowest level, centroids distributed at a distance of $2D$ belong to the second level, etc. The local peak assumption outlined earlier essentially requires the second level or higher for a centroid to qualify for a potential location of place formation. It is also assumed that the value of one unit increase in accessibility is monetized as v and remains constant over time, and that a transportation link is constructed with a uniform design speed of S_h kilometer per hour on top of a trail, for a fixed cost of C per kilometer.

To examine the sensitivity of model outputs to land use distributions, two experiments are executed with different sets of initial land use inputs: in Experiment 1, jobs or workers are randomly allocated across land use cells; in Experiment 2 number of jobs in a cell declines exponentially at a rate of β_1 with increased distance from this cell to the center of the space, while number of workers increases exponentially at a rate of β_2. In both experiments, each worker is assumed to have exactly one job such that the total number of workers is assumed to equal that of jobs. Whether in Experiment 1 or in Experiment 2, the total number of jobs or workers is fixed such that a cell on average has Q workers and Q jobs. Table 11.1 lists the default values of coefficients and parameters used in the experiments.

11.4 Hypotheses

Now that the model is set up to simulate the diffusion of a transportation network coupled with place formation, it is employed to test hypotheses regarding the locational and temporal advantages of first mover places / links in the network. Imagine an extreme case in which initial land uses are highly concentrated: pivotal locations where settlements are concentrated likely get established first; then transportation facilities are built to connect these places and become strategic routes with an expected high volume of through traffic. At this stage, it is posited that earlier established places and transportation facilities would gain FMA just because they acquired the best locations.

As the network spreads and connect to smaller places, more traffic is brought to earlier established places and strategic links. Therefore one would expect that FMA in the network would be reinforced as they benefit from the positive network effect during the evolutionary process of network growth.

The above advantages of first movers would be compromised, however, with:

Table 11.1 Model parameters and their specified values

Para.	Value	Unit	Description
θ	0.05	/min	Decay factor in trip distribution and the friction factor in node formation
μ	1		Relative value of accessibility to workers compared to accessibility to jobs
η	3		Scaling factor in node formation model
β_1	0.15	/km	Decay rate of jobs from region center
β_2	0.05	/km	Increase rate of workers from region center
C	1×10^6	\$/km	Construction cost of paved roads
D	5	km	Distance between adjacent land use centroids
Q	500	N.A.	Average number of jobs or workers in a land use cell
S_l	10	km/hr	Specified uniform speed of primitive trails
S_h	30	km/hr	Specified uniform speed of transportation links
v	0.05	\$/unit	Monetary value of a unit of accessibility to jobs

1. a less concentrated distribution of initial land use (the locations that earlier established places occupy are less attractive when desired land use activities are more evenly distributed over space), or
2. an over-invested network (if multiple routes are built between the same origin and destination, routes may compete with each other for the travel demand, thereby making earlier deployed routes less dominant).

Based on these speculations, the following hypotheses are proposed and tested in simulation experiments:

- *Hypothesis 1*: Earlier established places and transportation links gain FMA in a network, which will be reinforced as the network grows over time.
- *Hypothesis 2*: FMA is less evident in Experiment 1 than in Experiment 2, as the latter represents a higher concentration of initial land use.
- *Hypothesis 3*: FMA is less evident in a more redundant network as it indicates more intensive competition between parallel routes.

In order to test the hypotheses, measurement of first mover advantages in a transportation network is defined as follows. At the end of each inner-loop iteration, the model sorts established places and transportation links, respectively, by their formation times,[6] as well as by the current volumes of traffic flows that enter each place or link. If FMA does exist, earlier established places/links should attract more traffic.

---

[6] One and only one place is established per outer-loop iteration, so its formation time is indicated by iteration number. The formation times of links are also distinguished by their order in the sequence of construction in an inner-loop iteration. For instance, a link formation time labeled as "16.02" indicates this link is the second link that is built in Iteration 16.

The relationship between the ranks of places/links in terms of their formation times and those in term of their traffic volumes is examined by the Spearman rank order correlation test (Higgins, 2003), a non-parametric measure of correlation which we have used in the previous analyses. A negative Spearman correlation coefficient would indicate the presence of FMA, suggesting that the earlier a place / link is established, the larger volume of traffic it attracts, while a positive sign would suggest a disadvantage of first movers. The absolute value of the correlation coefficient would indicate the significance of FMA.

Two topological measures we have defined and utilized in the previous chapter (see Section 9.3) are calculated here to examine the correlation between FMA and the redundancy of a network. One of them is the gamma index (γ), a connectivity measure that quantifies the interconnection of nodes in a network by comparing the actual number of links with the maximum number of possible links in the network. The other measure of "circuitness" ($\phi_{circuit}$) evaluates the relative significance of circuit links in a network.

11.5 Results

Experiment 1 stopped at the 29th iteration while Experiment 2 at the 27th iteration. Figures 11.2 and 11.3 display the snapshots of the evolving network in the two experiments, respectively. Note that dots in gray represent the centroids of land use cells, some of which turn into magenta when established as places. The relative size of a dot indicates the agglomeration scale of land use activities (workers plus jobs) at a specific location. Edges in gray represent primitive trails, some of which turn into blue when built as transportation links.

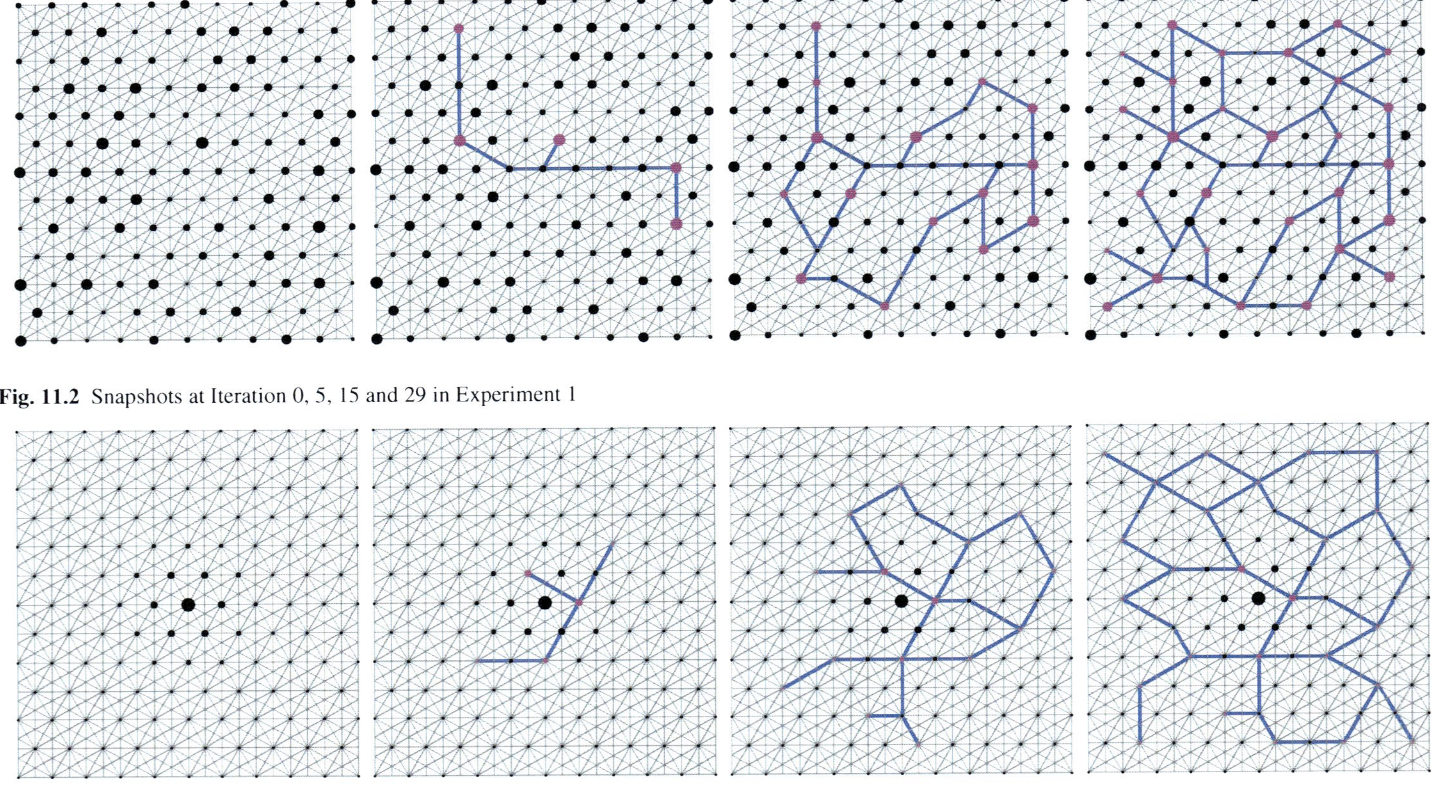

Fig. 11.2 Snapshots at Iteration 0, 5, 15 and 29 in Experiment 1

Fig. 11.3 Snapshots at Iteration 0, 5, 15 and 27 in Experiment 2

Spearman correlation tests were carried out for both places and transportation links at the end of every other iteration, and proposed topological measures were computed as well. The fluctuations of correlation coefficients and topological measures over iterations are displayed in Figures 11.4 and 11.5, respectively. Only the correlation coefficients with a 90 percent or higher confidence level (i.e., $p < 0.10$) are depicted.

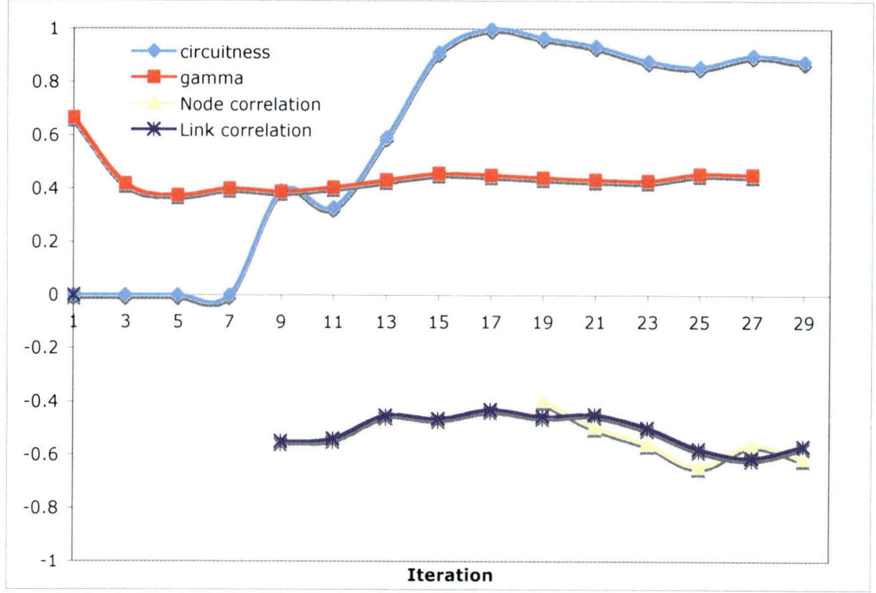

Fig. 11.4 Temporal change of topological attributes and correlation coefficients in Experiment 1

As can be seen, on most occasions both places and links exhibit a negative correlation between their formation times and traffic volumes, suggesting that the earlier a place or a link is established, the larger volume of traffic it attracts. This provides evidence for the existence of FMA in the deployment of a network. The general trend of increase in the absolute value of the correlation coefficient for both places and links is also observed, suggesting that FMA is reinforced over time as the network expands.

Starting with a more concentrated bell-shaped distribution of land uses, Experiment 2 results in stronger negative Spearman correlations relative to Experiment 1, suggesting a more concentrated distribution of land uses led to more significant advantages of first movers in the formation of a transportation network serving these land uses.

Both topological measures indicate the generally increasing redundancy of the simulated network. The fluctuation of the circuitness measure ($\phi_{circuit}$) is more volatile as compared to that of the gamma index (γ). The rises on the $\phi_{circuit}$ curve

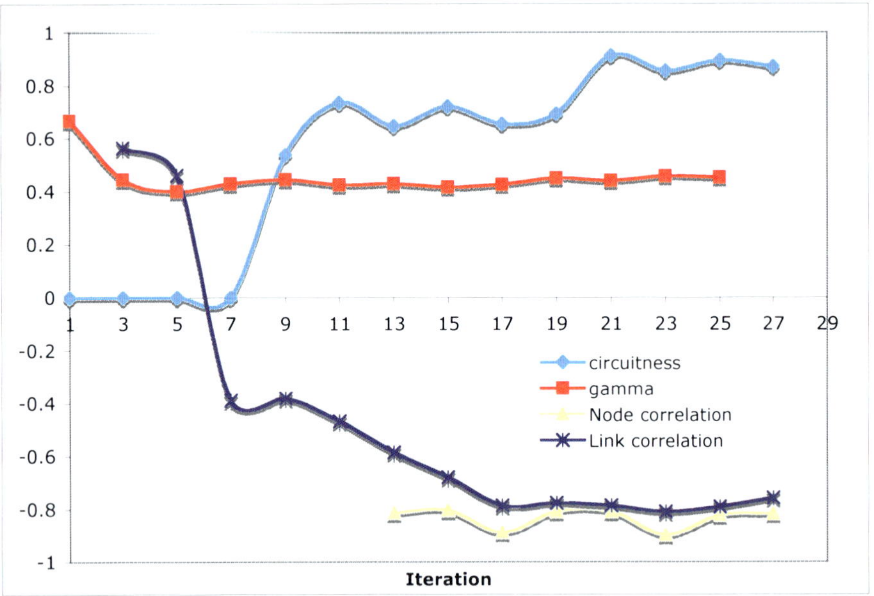

Fig. 11.5 Temporal change of topological attributes and correlation coefficients in Experiment 2

indicate the additions of circuit links that create alternative routes while the dips reflect the additions of branch links. Interestingly, as can be seen between Iteration 11 and Iteration 17 in Experiment 1 and Iterations 7-9 and 17-21 in Experiment 2, the increase in $\phi_{circuit}$ for the network is always accompanied by the lessening of the Spearman correlation. This observation suggests the inherent correlation between the strength of FMA and network redundancy as posited in the third hypothesis, although rigorous statistical tests are still in order to substantiate this relationship.

11.6 Sensitivity analysis

The values of model parameters listed in Table 11.1 are arbitrarily specified. To test the sensitivity of our analysis to these parameters, simulation was re-executed in a series of model runs with altered values for each parameter. The results are summarized in Table 11.2.

A smaller decay factor θ in the gravity model indicates a smaller impedance across a network and a higher level of accessibility. As can be seen, a smaller decay factor (0.02) in Run 1 for Experiment 1 resulted in much smaller Spearman correlation coefficients for both places (-0.482) and links (-0.209), indicating less prominent first mover advantages. This agrees with the speculation that advantages

Table 11.2 Spearman correlation coefficients in sensitivity analysis

		Value		Node		Link	
Run	Para.	Previous	Current	Coeff.	P	Coeff.	P
1	θ	0.05	0.02	-0.482	0.018	-0.209	0.016
2	β_1	0.15	0.1	-0.760	0.000	-0.654	0.000
3	C	1×10^6	5×10^6	-0.389	0.047	-0.611	0.000
4	Q	500	1000	-0.609	0.002	-0.581	0.000
5	S_h	30	60	0.710	0.000	-0.733	0.000

of first movers deriving from locational advantage in a network will be undermined as locational differentiation is lessened by reduction of travel impedance.

A smaller scaling factor η allows more randomness in the formation of places, thereby counteracting the first mover advantages. Similarly, a smaller value of μ or β_2 (in the bell-shaped distribution of land uses), specifying a lower concentration of initial land uses, is expected to lead to less evident FMA as well. To test, Experiment 2 was re-run in Run 2 with a different value of β_2 (0.10), and a weaker (and statistically significant) correlation is observed both for places (-0.760) and links (-0.654).

A lower value for accessibility (v) or a higher construction cost rate (C) leads to less construction in general, as the link formation process considers both benefit and cost. As the result of less network redundancy, more evident FMA is expected to be observed. Experiment 1 was re-run in Run 3 with a different value of $C(5 \times 10^6)$, and a stronger FMA for links (-0.611) was observed.

The distance between adjacent centroids D indicates the magnitude of the space and network, while Q indicates the scale of land use agglomerations. Changing either variable with the other remaining equal would change land uses and travel needs. Experiment 1 was re-run in Run 4 with a different value of Q (1,000). The resulting Spearman correlation coefficient for places equals -0.609, and that for links equals -0.581, indicating a slightly weaker FMA for places and a slightly stronger FMA for links.

The higher the design speed for transportation links is, the faster one can travel across established transportation links versus primitive trails, and a more evident FMA is expected in the more differentiated network. Re-running Experiment 1 with a higher speed (60), as expected, resulted in a much stronger correlation for both places (-0.710) and links (-0.733).

11.7 Discussion

In complement to the empirical illustrations in Chapter 7, this research took an *ex ante* approach to define and analyze first mover advantages in an abstract, controlled environment. Simulating first mover advantages in a transportation network involved several characteristics of the network.

First, are we considering nodes or links? (This research examines both). Nodes can connect to many links, links connect to only two nodes, so we expect their effects to differ. The capacity of nodes and links may be considered in different ways. Nodes may have limits on number of vehicles (flow) or on number of incoming or outgoing links (capacity). Similarly links may also have a flow-defined capacity limit, or it may be limited in the number of lanes. Since nodes can connect to more links than links can connect to nodes, we expect nodes to be more eligible for FMA than links.

Second, is there a preference for attaching to existing network elements in a particular way? Nodes may benefit from preferential attachment (Newman, 2001), while links benefit from preferential reinforcement (Yerra and Levinson, 2005), where existing links with large capacity attract more investment. There are both supply and demand related reasons for these preferences. Supply-related causes include economies of scale, economies of density, and lack of capacity constraint. Demand-related causes include network effects. Preferential attachment favors FMA.

Third, are we considering capacity constrained or capacity unconstrained networks? (This analysis considers unconstrained networks) All networks are ultimately constrained, but if we are dealing within the range that is for practical purposes unconstrained we get different answers than dealing in the range where the network is congested. Unconstrained networks are more likely to exhibit FMA.

Fourth, are there network externalities? When there exist network externalities, there is an advantage to hubbing. However, resulting from approaching capacity constraints, congestion externalities present a disadvantage to hubbing. How these net out depend on the technological characteristics of the mode as well as demand conditions. If hubbing benefits exceed congestion costs, then there exists a possibility for FMA. The London Underground and the hypothetical uncongested road network both illustrate FMA in transportation networks. The international system of airports do not possess FMA (the first airports do not carry more traffic than later airports). The international system of ports also do not possess FMA. However, the location of hub cities within an airline system is persistent. Airlines maintain hubs in the cities that they first established.

Fifth, are coordination advantages spatial, temporal, or both? Fixed infrastructure has spatial coordination, while transportation services (carriers such as airlines, shippers, buses, etc.) have both spatial and temporal coordination, and so the potential for coordination economies are greater. To illustrate, for a road network, the greatest spatial improvement (distance reduction) over a standard grid is circuity, which is on the order of 20 percent distance savings for a true air-line connection rather than a more typical network connection (Levinson and El-Geneidy, 2009). Speeds may change as well though.

For a carrier network, with scheduled services, hubbing can reduce schedule delays significantly by concentrating sufficient demand. Because the network economies are greater at hubs, the first hub, particularly if it is served by multiple parties, has a greater advantage over later hubs.

11.8 Findings and concluding remarks

This chapter investigates the existence and extent of first mover advantages in the deployment of spatial surface transportation networks using an *ex ante* approach. Based on empirical findings from Chapter 7 this chapter constructed the SONIC/PF model to replicate the incremental deployment of a transportation network coupled with place formation, and tested if earlier established places and transportation facilities gain locational advantages and if the advantages remain or change during the evolutionary process of network growth. Using traffic flow as a proxy of locational advantage in the early deployment phase of a network, Spearman rank order correlation tests revealed that the earlier a place forms or a link is built, the larger volume of traffic it attracts, and that the correlation becomes increasingly evident as the network grows, suggesting that first mover advantages not only exist in a transportation network, but also exhibit a reinforced strength during network diffusion. Simulation results also revealed that the extent of first mover advantages in a network correlates with initial land use distribution as well as network redundancy.

Different from the game-theoretic methods widely adopted in industrial economic studies, this research contributed to the literature by introducing a modeling approach, by which first mover advantage is defined and analyzed in a controlled environment. This study, though sacrificing some important considerations on land use development, congestion, ownership, and investment decision-making processes, keeps the model simple to focus on the particular question of first mover advantages. Simulation evidence has revealed the existence of first mover advantages in the deployment of surface transportation networks.

This research has important implications for strategic transportation planning, investment, and network design. The builders of transportation networks need to be exceedingly careful that the networks are appropriately sized and sited, since that will shape the use of those networks profoundly as the system adapts and locks-in. In addition, there are subsequent research questions yet to answer: How could economic and political initiatives factor in the deployment of a transportation system? How will transportation funds be allocated between existing infrastructure and new construction to facilitate the growth of a region? How could a transportation facility be appropriately sized and sited, not necessarily being optimal at the time, but being able to improve the system as a whole, considering subsequent construction in the future? Some of these questions may not be able to find answers without a more sophisticated and fully calibrated network model. This study, however, serves as a starting point in terms that it recognized the existence of first mover advantage and proposed a network growth model to investigate its contributing factors, which has the potential to be used as a planning tool in urban studies with the subsequent effects of plans or designs taken into consideration.

Chapter 12
Coevolution of Network and Land Use

12.1 Introduction

As American cities evolved in the first half of the 20th century, we saw a concentration of activities and development at the centers of cities. Since freeways were constructed from the 1960s, roads also became more differentiated with regard to their design speeds and capacities (certainly in the pre-auto era most unpaved streets were equally slow, with paved streets and highways and then freeways, some roads got much faster and carry more traffic). Faster roads have enabled decentralization of activities, and consequently led to a flattening of the density gradient of land use as centers of cities became relatively less important. In response to this trend, the representation of urban spatial structure in urban economics has migrated from the mono-centric models to the poly-centric models (Mori, 2006). Figure 12.1 displays the schematic representation of mono-centric versus poly-centric cities with primary and secondary traffic patterns (arrows) between activity centers (circles).

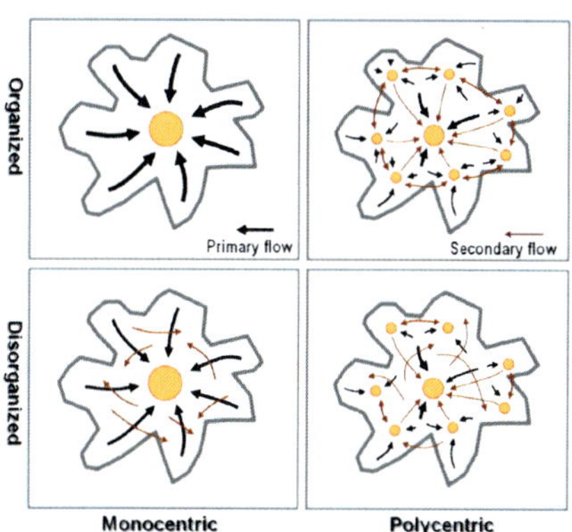

Fig. 12.1 Monocentric versus polycentric cities. Source: Rodrigue et al. (2006); adapted from Bertaud (2001).

This chapter examines the research question how differentiation of an urban road network relates to the concentration (or flattening) of surrounding land use activities. To answer this question unavoidably requires an evolutionary view. As widely recognized, transportation and land use are interdependent shapers of urban form. First, changes in land use alter travel demand patterns, which determine traffic flows on transportation infrastructure. Second, changed traffic flows on a daily basis drive the improvement of transportation facilities in the long run. Third, new transportation facilities change the accessibility pattern, which drives the re-location of activities and land use. During this process, both transportation and land use are evolving constantly, leading to salient spatial transformations such as agglomeration and centralization over space and transportation networks.

Although the concept of accessibility connects transportation with land use development, the change of transportation networks has seldom been considered in previous integrated land use-transport models (an overview of the literature on integrated land use-transport modeling is provided in Chapter 2). A possible explanation is that these models are already complicated enough. They usually involve multiple modeling approaches, incorporate numerous constraints and assumptions, and are estimated from empirical data, unavoidably leading to a comprehensive modeling framework including a wide variety of components. These models are so specific and complex that (a) they are difficult to replicate; (b) the relationships between components are entangled and implicit; (c) the emergent large-scale patterns in space and network are difficult to recognize and analyze. Lee (1973) also has an important critique.

In contrast to those complicated and all-encompassing models that do not provide an explicit perspective, this chapter approaches the research question by modeling the integrated dynamics of land use and roads in as simple a way as possible that captures salient properties, thereby enabling us to display and analyze the emergent hierarchy and agglomeration patterns of space and network on a large scale, as well as observe the interactions (reinforcement or counteraction) between the dynamics of roads and the development of land use. The specific simplifications and assumptions made in our model specifications will be discussed later.

Extending Chapters 8-10, this study models the co-evolution of land use and road network as a bottom-up, rather than a top-down process, by which interdependent location decisions of businesses (equivalently referred to as employment or jobs in this analysis) and residents (also called population or housing or workers or resident workers) are incorporated, as well as investment decisions of autonomous roads based on predicted traffic on a network. Planners and engineers would argue that while market-based land use may be constrained by zoning and plans, transportation network investments are decisions that are now driven, or coordinated, by centralized organizations such as state departments of transportation or metropolitan planning organizations that make major investment decisions using a forecasting model and planning process to test and evaluate alternative scenarios. Local jurisdictions, of which there are many in some metropolitan areas, make investments on lower level roads. Certainly these organizations do affect new investment, but the decision to build or expand a link is also constrained by many facts on the ground, actual traffic on the link, competing parallel links, and complementary and upstream and downstream links, the costs of expansion, and limited budgets (Levinson and Karamalaputi, 2003*a,b*). According to Krugman (1996), a self-organizing system of urban space and network will evolve into order and pattern, even based on simple, myopic, decentralized decisions of individual businesses and workers. If we can generate convincing collective representations of land use and network structure without any centralized planning or direction, perhaps planning is not as important in shaping urban areas as it is sometimes credited.

Using such a co-evolutionary model of transportation and land use, this chapter will examine the degree to which the dynamics of a road network and that of surrounding land use is reinforcing or counteracting each other. By this we ask whether a more hierarchical distribution of activities leads to a more or less hierarchical road network? Observation of historical evidence does not lead to a clear conclusion, as the development of a hierarchy of transit systems during the streetcar/subway era was accompanied with a concentration of development (especially employment) in the center of cities (from an undeveloped state), while the development of a hierarchical road network (from an underdeveloped and largely undifferentiated street system) occurred when those same cities were decentralized from a highly developed state. On the other hand, we also ask if the dynamics of land use reinforces or counteracts the differentiation of road networks. This research aims to examine these questions in a simulation environment with controlled initial conditions and quantitative measurements of spatial concentration.

The remainder of this chapter is organized as follows: the next section presents an introduction to the model developed for this study, which is followed by experiments and sensitivity analyses. The conclusions summarize the findings and suggest future directions for research.

12.2 Model

A new model, System of Integrated Growth of Networks And Land-use (SIGNAL), is developed in this study to represent the co-evolution of land use and transportation networks. An overview and inter-connection of this model is illustrated in Figure 12.2. The components of the model include travel demand, investment, accessibility, and land use.

12.2.1 Travel demand models

Similar to the SONG model, travel demand model converts land use data into traffic using a pre-existing network topology and determines the link flows by following the traditional planning steps of trip generation, trip distribution, and traffic assignment (for simplicity, a single mode is assumed). The travel demand model in SIG-

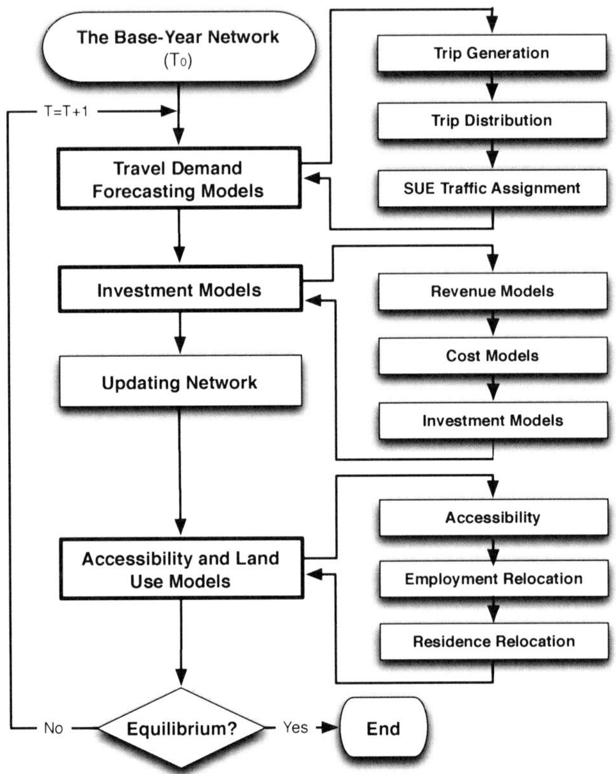

Fig. 12.2 Model framework of SIGNAL: System of Integrated Growth of Networks And Land-use

NAL differs from those in the previous chapters in two ways. First, rather than specify number of trips generated from and attracted to each land use cell/zone, this model associates trip production and attraction with two land use activities (employment and population) in a zone. Second, this model introduces Stochastic User Equilibrium (SUE) traffic assignment to predict travelers' route choices on a network that takes into consideration capacity constraints and congestion effects.

A simplified trip generation model estimates the number of vehicle trips that originate in or are destined for a zone as a linear combination of the quantities of employment (or jobs, denoted as J) and population (or workers, denoted as W) in this zone, without distinguishing trips by purpose:

$$O_i = \xi_0 + \xi_1 J_i + \xi_2 W_i \tag{12.1}$$

$$D_i = \psi_0 + \psi_1 J_i + \psi_2 W_i \tag{12.2}$$

Where O_i and D_i represent the number of trips that originate in or are destined for Zone i respectively.

A doubly constrained gravity-based trip distribution model is adopted to match both trip generation and attraction of locations based on a negative exponential function that assumes the interactions of zones decreases with the travel time between them:

$$N_{ij} = K_i K_j O_i D_j e^{\theta t_{ij}} \tag{12.3}$$

Where N_{ij} is number of trips from zone i to zone j; K_i, K_j are balancing coefficients. The parameters in this trip distribution model have been calibrated using the empirical data in the Twin Cities (details will be explained later in Chapter 15).

The variable t_{ij} denotes the generalized travel cost from zone i to zone j, which is calculated as:

$$t_{ij} = \begin{cases} \sum_a (\delta_{i,j}^a t_a) + t_{m,i} + t_{m,j} & i \neq j \\ t_{m,i} & i = j \end{cases} \tag{12.4}$$

Where $t_{m,i}$ and $t_{m,j}$ represent the generalized intra-zonal travel time in zone i and zone j, respectively. The generalized intra-zonal travel time captures a variety of costs incurred on trips that rise with land use intensity. It represents things like higher congestion levels, longer elevator waits in taller buildings, and greater difficulty of finding and undertaking parking that add to local travel time and travel cost in both zones. An intra-zonal generalized time penalty acts as a surrogate for all of the above. Assuming a simple quadratic relationship between the generalized intra-zonal travel time and land use density in zone i, $t_{m,i}$ is calculated as:

$$t_{m,i} = t_m^0 \left[1 + \left(\frac{P_i}{\bar{P}} \right)^2 \right] \tag{12.5}$$

Where t_m^0 is a specified base intra-zonal travel cost for all zones, P_i is the number of activities in zone i, while $\bar{P}$ represents the average number of activities across all zones. In our case,

$$P_i = J_i + W_i \tag{12.6}$$

The inter-zonal travel cost between zone i and zone j is computed as a summation of link travel cost along the shortest path from zone i to zone j. Dijkstra's Algorithm (Chachra et al., 1979) finds the shortest path from each node to all other nodes of the network. The generalized cost (t_a) on link a is calculated by adding monetary cost (toll) (with an appropriate conversion factor) to the actual travel time on this link (tolls are charged by the road agent):

$$t_a = \frac{l_a}{v_a} + \frac{R_a/\eta}{f_a} \tag{12.7}$$

Where l_a, v_a, f_a, and R_a respectively represent the length, average speed, traffic flow, and collected revenue of link a in a given time period. The parameter η represents the average value of time. The calculation of R_a will be discussed later.

A Stochastic User Equilibrium (SUE) algorithm is adopted in traffic assignment to predict route choices on a network according to perceived travel time, implementing Dial's Algorithm and Method of Successive Average (MSA) (Sheffi, 1985; Davis and Sanderson, 2002). Traffic assignment in a time period starts with the congested travel time resulting from the preceding time period, which makes the convergence in MSA much faster. The convergence rule in MSA specifies a maximal allowable link flow change equal to 0.5 (or a maximum of 100 iterations). A smaller maximal allowable flow change will result in a flow pattern that is closer to the equilibrium, but there is tradeoff between the accuracy and run time. The parameters in in the traffic assignment model have also been calibrated using the empirical data from the Twin Cities in Chapter 15.

12.2.2 Investment models

Investment models are similar in nature to those presented in the SONG model, except that independent links invest in capacity rather than their speed, assuming each link in a network is managed by an independent local agent.

A revenue model determines the toll a link collects during a given time period, depending on the traffic that uses this link. In a directed network, a link agent ab represents two directional arcs (link a and link b) on opposite directions. Let f_a and f_b respectively represent the flow traversing link a and link b for a given time period, the total revenue collected on both links by the agent can be calculated as:

$$R_{a+b} = \tau l_a (f_a + f_b) \tag{12.8}$$

Where τ is the regulated toll rate. A regulated toll rate across all the links simulates a distance based tax, which is the most common practice throughout the United States. Both link a and link b have the same length: $l_a = l_b$.

The cost to maintain links in their present usable conditions depends on link length, flow and capacity. The overall spending of the agent operating links a and b is calculated as:

$$S_{a+b} = l_a v_{f,a}^{\sigma_2} (f_a^{\sigma_1} + f_b^{\sigma_1}) \tag{12.9}$$

Where the coefficients σ_1 and σ_2 are specified flow and capacity powers in the equation.

An investment model assumes each agent spends all its available revenue at the end of a time period myopically, without saving it for the future. If the revenue exceeds the maintenance cost, remaining revenue will be invested to expand the capacity of subordinate links. In contrast, if the revenue is insufficient to cover the cost, road conditions will deteriorate and link capacity will drop until the link is eventually abandoned. This investment policy adopted by each agent can be expressed in a simplistic form as:

$$C_a^{k+1} = C_a^k \left(\frac{R_{a+b}^k}{S_{a+b}^k} \right)^\rho \tag{12.10}$$

Where $C_a = C_b$ is the capacity of link a and b, which changes with iteration (k), respectively, while ρ is a specified coefficient that affects the speed of convergence. As implied by Equations 12.8-12.10 and specified parameters (detailed in Table 12.2), a network equilibrates when the flow on each link equals road capacity in quantity.

Zhang and Levinson (2005) estimated the relationship between the free flow speed of a link and its capacity in a log-linear model based on the empirical data in the Twin Cities. The log-linear relationship is adopted here to update the free flow speed (v_f) of a link after its capacity is changed:

$$v_{f,a} = \omega_1 + \omega_2 Ln\left(C_a\right) \tag{12.11}$$

where ω_1 and ω_2 are two coefficients in the log linear equation while C_a is the capacity of link a.

The relationship between the free flow speed and congested speed of a link is defined by the BPR function (Bureau of Public Roads, 1964) as:

$$v_{c,a} = v_{f,a} \left[1 + \alpha \left(\frac{f_a}{C_a} \right)^\beta \right] \tag{12.12}$$

where α and β are the coefficients of the function, assumed to equal 0.15 and 4.0, respectively.

12.2.3 Accessibility and land use models

Accessibility reflects the desirability of a place by calculating the opportunities and activities which are available from this place via a transportation network but are also impeded by the travel cost on the network. Suppose an urban space is divided into a set of Traffic Analysis Zones (TAZs) or land use cells that contain both employment (jobs) and population (workers). The accessibility in each cell (to employment and population) is computed respectively using a negative exponential measure:

$$A_{i,J} = \sum_m J_m e^{-\theta \, t_{im}} \tag{12.13}$$

$$A_{i,W} = \sum_m W_m e^{-\theta \, t_{im}} \tag{12.14}$$

where $A_{i,J}$ is the accessibility to employment (jobs) from zone i while $A_{i,W}$ is the accessibility to population (workers). The coefficient θ indicates how the accessibility of a zone declines with the increase of travel time to the zone. This coefficient is also used in Equation 12.3 to indicate the travel time impedance factor.

A land use model is then developed to reflect how the distribution of jobs and resident workers responds to changes in the accessibility patterns, while keeping the total number of jobs and the total number of resident workers as constant. The land use model is simplified in the sense that accessibility to jobs and accessibility to resident workers are the only factors that affect the locational decision made by jobs and resident workers. As accessibility is essential in the relationship between transportation and land use, other factors such as land price and administrative policies are excluded to keep this relationship succinct and clear, thus enabling simple accessibility-based rules to which independent location choices can be made. To be representative, our land use model contains both centripetal and centrifugal forces, that is, a force of attraction (e.g. economies of agglomeration) and a force of repulsion (a desire on the workers part for spatial separation, keeping all activities from locating at a single point). We assume resident workers want to live near jobs, but far from other resident workers (to maximize available space and to avoid potential competitors for employment), while jobs want to be accessible both to other jobs and to resident workers (who are their suppliers of labor and customers). The following stylized models are developed to track the dynamics of land use based on independent decisions of jobs and resident workers with regard to their locations.

Equations 12.15 and 12.16 estimate the desirability (potential) of a zone to attract jobs and resident workers, respectively.

$$U_{i,J} = A_{i,J}^{\lambda_1} A_{i,W}^{\lambda_2} \tag{12.15}$$

$$U_{i,W} = A_{i,J}^{\lambda_3} A_{i,W}^{\lambda_4} \tag{12.16}$$

where λ_1-λ_4 are coefficients that indicate the positive or negative relationship between accessibility and land use. Note that both accessibility to jobs and accessibility to resident workers reinforce the desirability of a zone for jobs, implying only a centripetal force exists in shaping the distribution of jobs (though intrazonal transportation costs do increase with density). On the other hand, accessibility to jobs reinforces while accessibility to workers counteracts the residence desirability of a zone. Thus we have $\lambda_1, \lambda_2, \lambda_3 > 0$ and $\lambda_4 < 0$. An empirical study by El-Geneidy and Levinson (2006) corroborates our assumptions. Based on 44,429 home sale records for the year of 2004 in the Twin Cities metropolitan region, a hedonic model discloses the connection between single-family residence property values and accessibility to jobs and resident workers, with other factors controlled. Accessibility to jobs did show a statistically significant positive effect on home sale values, while accessibility to resident workers did show a statistically significant negative effect. Furthermore, their coefficients are approximately equal in the model (though having opposite signs), implying the centripetal force and the centrifugal force on residence location are equally strong. Based on their findings, this study further assumes $\lambda_3 = -\lambda_4$.

Jobs and resident workers are then reallocated across zones at time period $k+1$ according to the following equations, basically in proportion to the desirability of each zone at the preceding time period k, except for the parameter μ introduced to indicate the reluctance for jobs and workers to move away from the original location. For simplicity, the totals of jobs and resident workers are held constant over time.

$$J_i^{k+1} = \sum_m \{ J_m^k \frac{(U_{i,J}^k)^\mu}{\sum_n (U_{n,J}^k)^\mu} \} \tag{12.17}$$

$$W_i^{k+1} = \sum_m \{ P_m^k \frac{(U_{i,W}^k)^\mu}{\sum_n (U_{n,W}^k)^\mu} \} \tag{12.18}$$

where:

$$\mu = \begin{cases} 1, \; if \; i = j \\ < 1, \; if \; i \neq j \end{cases}$$

Figure 12.3 illustrates the feedback relationship between the network and land use variables within our system of co-evolution. An arrow with a plus (+) or minus (-) between two boxes shows a positive or negative relationship between the boxes. As can be seen, road expansion increases capacity, which improves free flow speed; the increased capacity increases cost, then forces the capacity back according to the investment rules. The improvement of travel time increases traffic flow, which increases the revenue and facilitates road expansion. The improvement of travel time also increases both accessibility to jobs and accessibility to houses. Employment density is positively associated with both accessibilities while population density is negatively impacted by accessibility to houses. Increased employment or population density increases intrazonal travel time, which offsets the improvement of travel time due to road investment.

After investing (or dis-investing) in each link in the network, computing accessibility, and relocating land use, the time period is incremented and the whole process is repeated. In this study one time period represents a hypothetical year as the day-to-day traffic on the network is predicted and converted to yearly traffic for road investment models.

12.3 Hypotheses and experiments

The model is set up to scrutinize two sets of hypotheses:

The first is the degree to which hierarchies of road networks are reinforced or counteracted by the dynamics of land use. It is posited that this depends on initial land use and network conditions. Initially flat road networks become more concen-

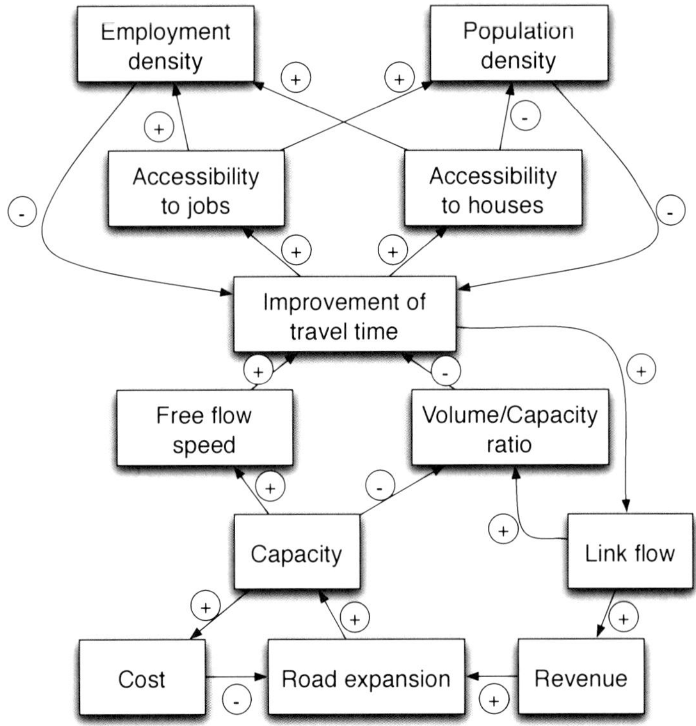

Fig. 12.3 The feedback relationship in the transportation/land use system

trated (*Hypothesis 1*), and initially concentrated networks become less so (*Hypothesis 2*) when land use are allowed to vary rather than remain constant. That is, they reinforce to a point, and counteract beyond some point.

Similarly, the second asks if the dynamics of transportation networks reinforces or counteracts concentration of land use. It is posited that initially flat land use gradients become more concentrated (*Hypothesis 3*), and initially concentrated land use becomes less so when road networks are allowed to vary rather than remain constant (*Hypothesis 4*).

As outlined in Table 12.1, three sets of experiments are conducted to test our hypotheses. The first fixes the land use, and explores how the network evolves in response to those fixed land use. The second fixes the road network, and explores how the land use evolves in response to the fixed network. The third allows both the network and the land use to evolve simultaneously.

Each set of experiments was tested under different initial conditions. As shown in Table 12.1, Experiments 1(a), 2(a), 3(a), and 3(c) specify a uniform network in which the same initial conditions are specified for all the links except for their locations: each link is 1 km in length with a free flow speed of 50 km/h, and a capacity of 800 veh/h. Experiments 1(b) and 3(b), on the other hand, assume a concentrated

Table 12.1 Specification of experiments

No.	Initial conditions			Dynamics	
	Link capacity	Employment	Population	Roads	Land use
1a	Uniform	Uniform	Uniform	Evolving	Fixed
1b	Concentrated	Uniform	–	Evolving	Fixed
2a	Uniform	Uniform	–	Fixed	Evolving
2b	Uniform	Concentrated	–	Fixed	Evolving
3a	Uniform	Uniform	–	Evolving	Evolving
3b	Concentrated	Uniform	–	Evolving	Evolving
3c	Uniform	Concentrated	–	Evolving	Evolving

distribution of network capacity by which capacity concentrates on the links located at the center of the network and decreases as the distance to the center increases. Experiments 1(a), 1(b), 2(a), 3(a), and 3(b) specify uniform land use with the quantities of both population and employment equal to 1,000 in each zone; while Experiments 2(b) and 3(c) assume a concentrated distribution of employment as number of jobs in a zone declines exponentially with the increase in the distance between this zone and the center of the study area.

With two quantitative measures the Gini index (G) and equivalent radius (r) defined in Section 9.3, we are able to quantify for each experiment the changes in spatial concentration going from the initial state to the equilibrated state, denoted by ΔG and Δr, respectively. Accordingly, the hypotheses can be tested by comparing the changes in the spatial concentration of networks or land use from different experiments.

Simulation experiments were conducted in a hypothetical metropolitan area where both the population and employment are distributed over a two-dimensional grid. For simplicity, the experiments here are conducted over a square planar surface, stretching 20 km in both dimensions, divided into a 20×20 grid lattice of land use cells (400 zones). Each zone occupies one square kilometer of land. Whether starting from a uniform distribution or a concentrated distribution, a total of 400,000 jobs are distributed over this city, which is equivalent to an average of 1,000 jobs in each zone. Total population equals 400,000 as well (such that each resident holds a job). Two-way roads connect the centroids of each pair of adjacent zones, thus forming a 20X20 grid of road network as well, comprising 400 nodes and 1,520 links.

Table 12.2 lists parameters and their values for our experiments. As explained in Table 12.2, the toll rate and value of time are adopted from empirical estimates; the slope coefficient that defines the log-linear relationship between link capacity and free flow speed is estimated by Zhang and Levinson (2005) using the empirical data in the Twin Cities, while the intercept coefficient is estimated so that the specified initial link speed (50 km/h) and capacity (800 veh/h) fit into this equation. Among those parameters that are arbitrarily specified for the models, some were tested in the experiments using sensitivity analysis, which will be discussed later.

Table 12.2 Model parameters and their specified values

Para.	Description	Value	Source
ξ_0, ξ_1, ξ_2	Trip production rates	0, 0.5 trips/psn, 1.0 trips/psn,	Specified
ψ_0, ψ_1, ψ_2	Trip attraction rates	0, 1.0 trips/psn, 0.5 trips/psn	Specified
t_m^0	Base intra-zonal travel time	10 min	Specified
η	Value of time	$10 /h	Empirical estimates
τ	Toll rate	$1.0/veh-km	Specified
σ_1, σ_2	Coefficients in cost model	0, 1	Specified
ρ	Capacity reduction factor	0.1	Specified
ω_1, ω_2	Coefficients in the capacity/free flow speed log-linear function	-30.6 km/hr, 9.8	Empirical estimates
α, β	Coefficients in BPR function	0.15, 4.0	Typical values
θ	Impedance factor in accessibility model	0.048/min	Empirical, calibrated
λ_1 - λ_4	Coefficients in desirability models	1.0,1.0,0.9,-0.9	Specified
μ	Reluctance to move	0.8	Specified

12.4 Results

12.4.1 Results related to Hypotheses 1 & 2

Allowing investment in road capacities while fixing the land use, Experiments 1(a) and 1(b) are similar in nature to those presented in Chapter 8, though differing in specific parameters, the route assignment model, and initial conditions. Experiments 3(a)-3(b), on the other hand, allow both land use and network to evolve, thus generating different network structures and land use patterns.

The evolving network patterns were analyzed by plotting the measures of the Gini index (G) and equivalent radius (r) for network capacity on a time horizon, shown in Plot (i) and Plot (ii) of Figure 12.4, respectively. Each plot displays four fluctuations for the experiments starting from a uniform vs. concentrated network with and without land use dynamics, that is, Experiments 1(a), 1(b), 3(a), and 3(b). For clarity data points are plotted every other iteration.

As indicated by the Gini index in Plot (i) and by the equivalent radius in Plot (ii), the networks in Experiments 1(a), 1(b), 3(a), and 3(b) all reached equilibrium within the first 80 iterations and remained unchanged thereafter. Starting from a uniform state with fixed land use (for the road network, $G=0.0$ and $r=8.0$ km), the network in Experiment 1(a) adjusted its distribution of road capacity to the contemporary traffic patterns and concentrated capacity additions where roads are most used. This led to a gradual increase in network concentration (indicated by a rising Gini index of roads in Plot i and a dropping radius in Plot ii) until it reached equilibrium (where $G=0.10$

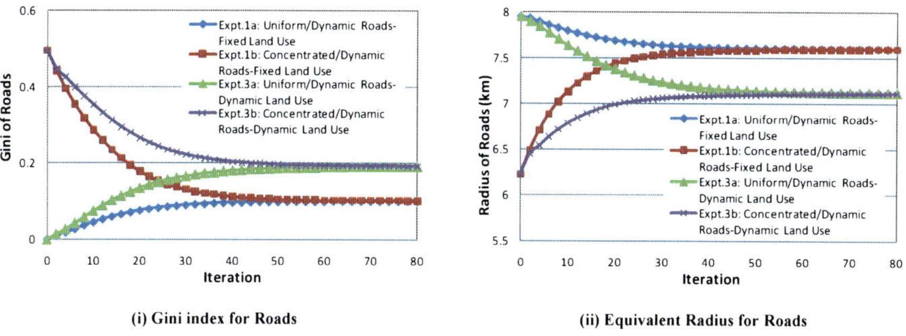

(i) Gini index for Roads   (ii) Equivalent Radius for Roads

Fig. 12.4 Measures of network patterns

and r=7.6 km). Starting from a concentrated state, on the other hand, the network in Experiment 1(b) saw a flattening of network capacity distribution, but reached the same level of concentration. More interestingly, Experiments 3(a) and 3(b), both allowing land use to evolve, converged on a higher level of network concentration (G=0.19 and r=7.1 km).

The fact that road networks converged to the same level of spatial concentration whether starting from a uniform or a concentrated state (all else being equal) suggests the existence of a stable hierarchical order in network dynamics regardless of initial conditions. Furthermore, the experiments allowing land use dynamics (3(a) and 3(b)) generated a consistently higher Gini index and lower radius compared to their counterparts with fixed land use, suggesting the evolution of land use distribution reinforces the differentiation of roads. These findings support our first two hypotheses that when land use is allowed to vary, initially flat road networks become more concentrated, and initially concentrated networks become less so.

Our findings can be further corroborated by the snapshots of emergent network patterns shown in Figure 12.5. Starting from the same uniform network in Plot (i), Plots (iii) and (v) display the two equilibrated networks with fixed vs. evolving land use from Experiments 1(a) and 3(a), respectively. Starting from the concentrated network in Plot (ii), Plot (iv) and Plot (vi) display the two equilibrated networks with fixed vs. evolving land use from 1(b) and 3(b). Different levels of capacity are displayed in five different colors and thickness in a relative scale. Obviously, the resulting networks with evolving land use in Plots (v) and (vi) are more concentrated than their counterparts with fixed land use, suggesting land use dynamics reinforces the hierarchical distribution of road infrastructure in the context of co-evolution of network and land use.

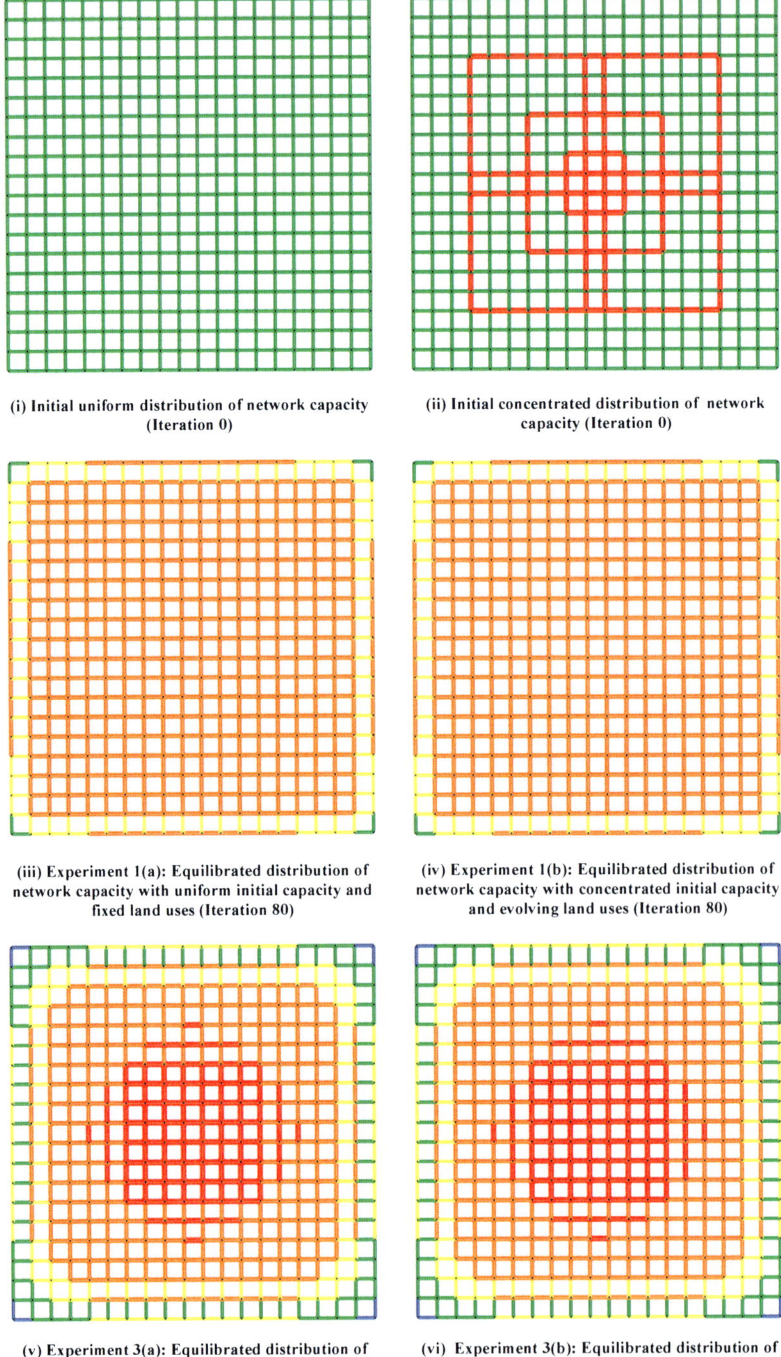

Fig. 12.5 Emergent network patterns

12.4.2 Results related to Hypotheses 3 &4

Plots (i)-(iv) in Figure 12.6 each displays four fluctuations for the experiments starting from a uniform vs. concentrated distribution of land use with and without road dynamics, that is, Experiments 2(a), 2(b), 3(a), and 3(c). Plots (i) and (ii) look at changes in the spatial distribution of employment, while Plots (iii) and (iv) examine changes in population.

Starting from uniformly distributed land use in Experiments 2(a) and 3(a), the Gini index of employment is constantly increasing (as shown in Plot i) while the radius is always dropping (Plot ii), showing a strong trend of agglomeration and centralization of employment. On the other hand, the Gini index is dropping while the radius is increasing in Experiments 2(b) and 3(c), showing the initially concentrated employment is spreading out over time. With 80 iterations the curves show a strong tendency of convergence. Similar to what we have observed in road dynamics, whether starting from a uniform or concentrated state, the distribution of employment tends to converge on the same level of concentration (Experiments 2(a) and 3(a) converge on a Gini index of 0.09 and a radius of 7.7 km, while 2(b) and 3(c) on a Gini index of 0.18 and a radius of 7.3 km), indicating a stable distribution of employment may emerge from different initial conditions. Experiments allowing road dynamics (3(a) and 3(c)) generated a consistently higher Gini index and lower radius than their counterparts with a fixed road network (2(a) and 3(b)), suggesting investment on roads (from an initial uniform state) reinforces the concentration and centralization of employment.

The evolution of population distribution works out differently. As shown in Plots (iii) and (iv) in Figure 12.6, starting from the same uniform distribution of land use in Experiments 2(a) and 3(a), the Gini index for population keeps increasing while the radius keeps dropping, showing that people are clustering and moving toward the region center. In Experiments 2(b) and 3(c), the distribution of population converged even faster than in 2(a) and 3(a), under the centripetal force coming from the initial concentration of jobs in the center. Finally Experiments 2(a) and 2(b), or Experiments 3(a) and 3(c) tend to converge on the same Gini index and equivalent radius of population, although they started from different distributions of employment.

The observation that initially flat land use becomes more concentrated while initially concentrated land use becomes less so, and they tend to converge on the same hierarchical distribution suggests a stable hierarchical distribution of land use may emerge from different initial conditions. Experimental results also show that the hierarchical distribution of land use is reinforced when the road network is allowed to vary rather than remain constant, indicated by a consistently higher Gini index and consistently lower equivalent radius in the former case, while Experiment 3(c) also demonstrates that the initial concentrated gradient of land use is flattened as roads are expanded from an undifferentiated and relatively slow road network, which mirrors development in the United States and other advanced countries during the twentieth century when the increasing differentiation and development of roads was accompanied with a flattening of the land use density gradient.

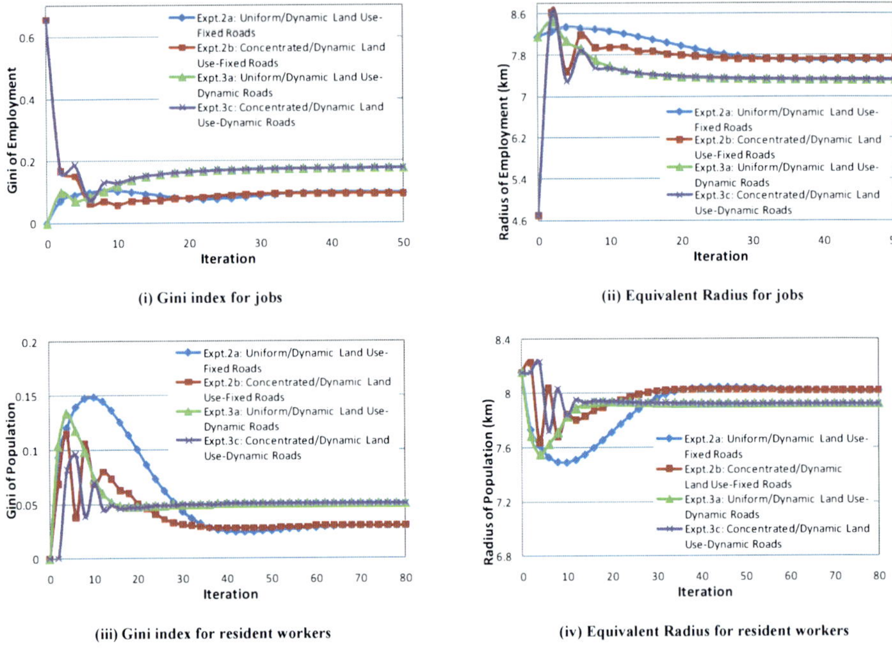

Fig. 12.6 Measures of network patterns

Snapshots of employment distribution patterns are visualized in Figure 12.7 (i)-(vi). According to the relative density of employment, zones are divided into 5 hierarchies and displayed in different colors. Plots (i) and (ii) in the figure display two different initial states of employment distributions, that is, flat and concentrated respectively. Plots (iii) and (v) display the respective hierarchical patterns emerging from the uniform state at the end of Experiments 2(a) and 3(a), while Plots (iv) and (vi) display the emergent patterns which are flattened from the initial concentrated state in Experiments 1(b) and 3(c), respectively. As can be seen Plot (iii) vs. (iv) and Plot (v) vs. (vi), similar distribution patterns have emerged from different initial states, indicating a stable hierarchical distribution of land use may emerge from the co-evolution of land use and transportation. Further, the employment distribution in Plot (v) and Plot (vi) with evolving roads appear more concentrated than their counterparts in Plots (iii) and (iv) which fixed the road network, further demonstrating road dynamics may reinforce the differentiation of land use.

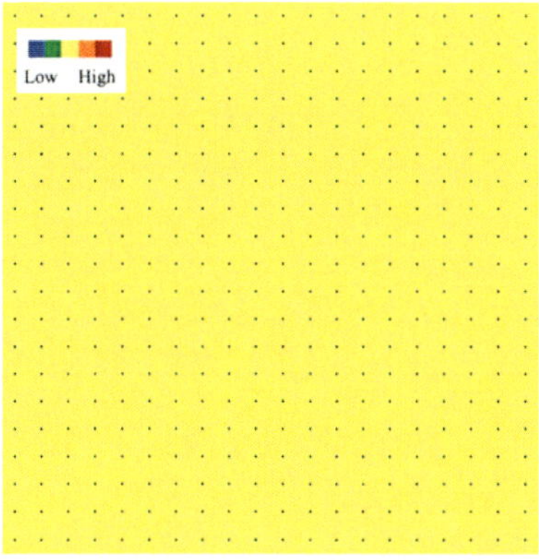

(i) Initial uniform distribution of employment
(Iteration 0)

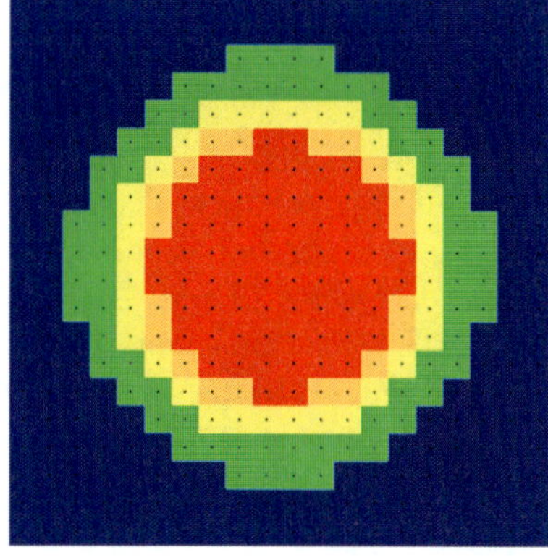

(ii) Initial concentrated distribution of employment
(Iteration 0)

(iii) Experiment 2(a): Equilibrated distribution of
employment with uniform initial distribution and fixed
roads (Iteration 80)

(iv) Experiment 2(b): Equilibrated distribution of
employment with concentrated initial distribution and
fixed roads (Iteration 80)

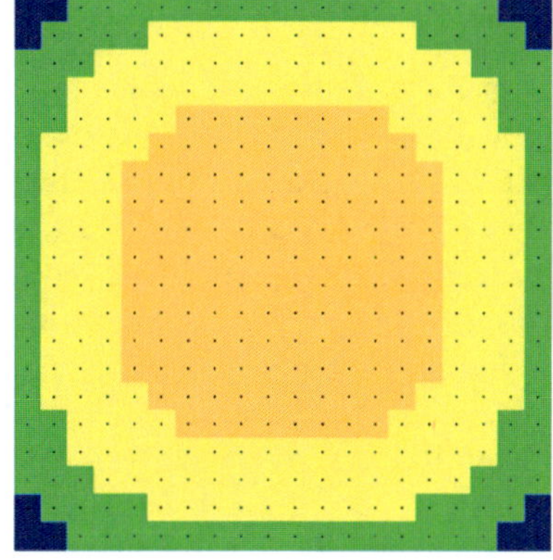

(v) Experiment 3(a): Equilibrated distribution of
employment with uniform initial distribution and
evolving roads (Iteration 80)

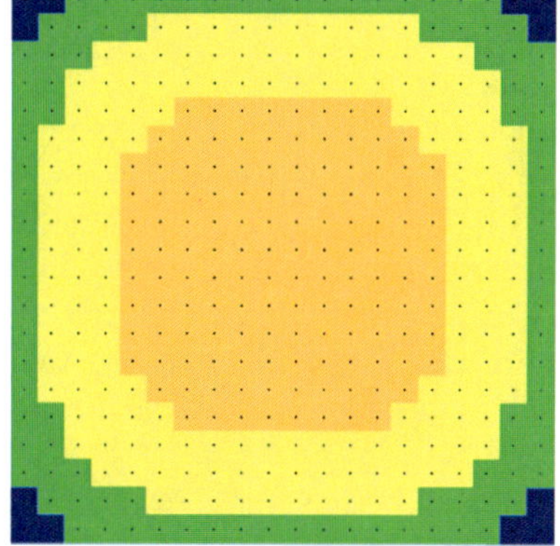

(vi) Experiment 3(c): Equilibrated distribution of
employment with concentrated initial distribution and
evolving roads (Iteration 80)

Fig. 12.7 Emergent employment patterns

12.5 Sensitivity analysis

Changing the specified values of the parameters in the models of road dynamics and land use dynamics may affect the emergent spatial patterns significantly. The sensitivity of these parameters is thus examined as follows.

For road dynamics, the specified flow power σ_1 and speed power σ_2 (0.0 and 1.0) in our road cost model simplify the road network dynamics model such that equilibrium is reached when the volume equals the capacity on each link. The parameter ρ in the investment model does not affect the equilibrium state, but it defines how fast road investment responds to road conditions, thus affecting the speed of reaching equilibrium.

The parameters in the land use model, including the decline factor θ, the coefficients in the zonal desirability model (λ_1-λ_4), and the reluctance factor in the land use reallocation model μ, may affect the emergent spatial patterns as well. The decline factor θ in trip distribution model and accessibility models determines how fast the attraction of a place declines with the increase of (generalized) travel cost to that place, reflecting the extent to which the change of travel time can affect the distribution of land use. The coefficients λ_1- λ_4 determine the power of centripetal and centrifugal forces in locational decisions with related to accessibility to jobs and accessibility to resident workers. The reluctance factor μ indicates jobs and resident workers' willingness to stay in the original location.

Different values for four parameters ρ, θ, λ_3,[1] and μ were tested in Experiment 3a with uniform initial conditions and allowing both network and land uses to vary, and the results are summarized in Figure 12.8. To be concise, only the Gini index of job distribution is plotted.

As can be seen in Plot (i), while the spatial distribution of network capacities evolved into the same pattern with different values of ρ, the speed of reaching equilibrium varies. A higher value of ρ indicates a faster convergence of road dynamics. As shown in Plot (ii), a smaller value of θ resulted in a higher level of employment concentration because it puts it puts more weight on travel time in the locational choices of workers. Plot (iii) displays the fluctuations of the Gini index with different values of λ_3. Note that λ_4 changes with λ_3 such that λ_4 is always equal to -λ_3. The increase in λ_3 reinforces the agglomeration, while the increase of λ_4 (in absolute value) enhances the centrifugal force. Jointly their changes did not affect the resultant spatial pattern significantly. Plot (iv) tests five different values of the reluctance factor μ. As can be seen, the resultant spatial pattern is very sensitive to this coefficient. Note that the change in μ affects the distribution of jobs and that of resident workers differently. With a higher value of μ (lower moving reluctance), the distribution of jobs will become more concentrated, while the distribution of resident workers could become less so (because of the reinforced centrifugal power). A less concentrated distribution of resident workers may in turn counteract the agglomeration of jobs. The sensitive analysis discloses that a higher value of μ counteracts the

---

[1] To be consistent with the empirical findings from El-Geneidy and Levinson (2006), λ_4 changes accordingly with λ_3 such that $\lambda_3 = -\lambda_4$ always holds.

agglomeration of jobs below a turning point of 0.85, and reinforces that above the turning point.

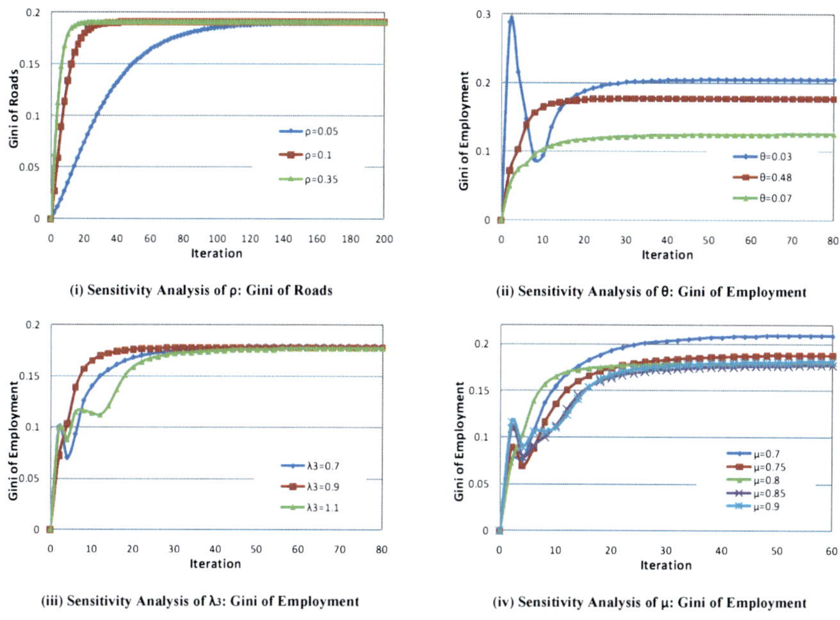

(i) Sensitivity Analysis of ρ: Gini of Roads

(ii) Sensitivity Analysis of θ: Gini of Employment

(iii) Sensitivity Analysis of λ3: Gini of Employment

(iv) Sensitivity Analysis of μ: Gini of Employment

Fig. 12.8 Sensitivity analyses for two parameters

12.6 Findings and concluding remarks

This study models the co-evolution of land use and transportation network as a bottom-up process by which the re-location of land use activities and expansion of roads are driven by interdependent decisions of individual businesses, workers, and road agents according to simple decision rules. The model was kept simple so that collective spatial patterns of land use distribution can be displayed and analyzed without multiple conflating factors, while the sensitivity of these patterns were also discussed. The Gini index and equivalent radius were calculated to trace the evolution of spatial patterns. Simulation experiments suggest that there may exist an inherently stable hierarchical distribution of network capacity such that flat networks become more concentrated and concentrated network become less so. Experimental results further demonstrated that the agglomeration and centralization of network infrastructure is reinforced by the dynamics of employment and population under the tension of pushing and pulling forces. Land use organization and concentration make the road network more concentrated than it otherwise would be. Similarly, simulation experiments suggest that there may exist an inherently stable hierarchical distribution of land use activities so that initially flat land use become more concentrated and concentrated land use become less so. It is also demonstrated that the agglomeration and centralization of employment and residence is reinforced by the dynamics of the underlying road network. Since these observations have been replicated in a spontaneous process based on completely decentralized decision-making, this reinforcement phenomenon is suggested to be an emergent property of the co-evolution of land use and road network.

Part V
GOVERNANCE AND PLANNING

This part explores alternative governance schemes in the provision of a transportation network from an evolutionary perspective. It also constructs network growth models for governance and forecasting studies from a normative view.

Chapter 13
Governance Choice - A Theoretical Analysis

13.1 Introduction

Policy debate has lasted over the past twenty years regarding the distribution of federal highway funding in the US Federal-Aid Highway Program. The controversy arose as the program's payment to the individual states did not match the amount of federal highway taxes each state's highway users pay to the Highway Trust Fund (HTF), which is commonly referred to as the "donor-donee" problem (Kirk, 2004). Donee states like Alaska experience high returns from the HTF, donor states like California contribute more than they receive, while some states like Ohio have experienced instability and cycling in terms of their donor/donee status.[1] In general, the allocation of federal highway funds appears to be based on formulas that are biased in favor of states with relatively low highway use to maintain national integration (Johnson and Libecap, 2000). In such an imbalanced allocative program, a fiscal correspondence problem arises when policy makers in donee states spend to benefit their constituent travelers and suppliers, but whose welfare costs are borne by taxpayers in all the states.

While decision-making at a higher government level in a multi-jurisdictional road system seems to be desirable in terms of better coordination of jurisdictional interests, it also brings institutional and political problems that may lead to inequity or economic inefficiency. The "donor-donee" problem discussed above is one of the examples. Regional transportation institutions have their mis-allocation problems too. As Haynes et al. (2005) pointed out, Metropolitan Planning Organizations (MPOs) usually adopt the "one government/one vote" decision structure regardless of the heterogeneous distribution of regional population and transportation demand, which tends to under-represent central cities and poorer populations, and over-represent suburbs and higher income groups. The legislative process at the federal level causes additional problems. Politicians form coalitions to maintain politi-

---

[1] As Kirk (2004) demonstrated with highway statistics, Ohio has been a donee in six fiscal years during the period of 1981-2002, and a net donor for the other years.

cal support and reach policy agreements in the Congress, at the obvious cost of the economic efficiency of their policy decisions.[2]

The institutional organization in transportation provision has been evolving over time. It has long been observed that the development of transportation systems has been characterized by "a constantly shifting mix of private enterprise, on the one hand, and government initiatives at local, state, and national levels, on the other hand" (Taaffe et al., 1996). The shift between alternative regulatory regimes has profoundly influenced the provision of transportation infrastructure, both in the policy-making process, infrastructure financing, and in the shaping of specific routes and network patterns. Narrowing our focus to the government provision of transportation infrastructure, although the roles of governments at different levels were always intertwined in reality, some centralizing tendency to move from local to state to federal levels has been noted when the major modes of waterways, railways, turnpikes, and roads were deployed in the US (Taylor, 2004; Garrison and Levinson, 2006; Bogart, 2005). Most US turnpikes, for instance, were local throughout the nineteenth century, initially organized by larger cities connecting ports to their hinterlands. As transportation systems developed, however, the roles of the state and federal government became more and more prominent. Railway consolidation and regulation, as well as a revival of inland waterways and certain canals would not have been possible without government involvement and financial aid at the state and federal levels. While the promotion of road transportation had been primarily at local and state levels during the early twentieth century, the federal government had taken over the major promotional role during the interstate highway era post-1956. Evidence is available in Figure 13.1, in which percentage of total receipts for highways by government levels are calculated based on the data from Federal Highway Administration (2002), and displayed for each year during 1945-2002. Over this period, 1956 witnessed an increase in the federal share of highway receipts, immediately following the establishment of the Federal Highway Trust Fund (HTF) to finance the construction of the Interstate highway system (McDaniel and Coley, 2004). As infrastructure has aged and congestion grown after the completion of the Interstate system, the call for better coordination between jurisdictions and departments involved in the management, operations, and expansion of regional road systems have been growing in intensity. In response to the interconnected and interdependent nature of current road systems, the establishment of Metropolitan Planning Organizations (MPOs) and various Regional Operating Organizations (ROOs) in recent years has marked a shift of institutional structure of transportation decision making from a local to a regional level (Haynes et al., 2005).

This chapter, with a particular focus on the government provision of road infrastructure, examines how transportation policy decisions have been instituted by various government levels, and which level of government would be preferable to deliver infrastructure projects at a particular stage of transportation development. From an evolutionary perspective, it also explores why the institutional structure of policy making has spontaneously shifted during the course of transportation devel-

---

[2] As Winston (2000) described, "transportation bills are loaded with demonstration or 'pork barrel' projects to ensure passage".

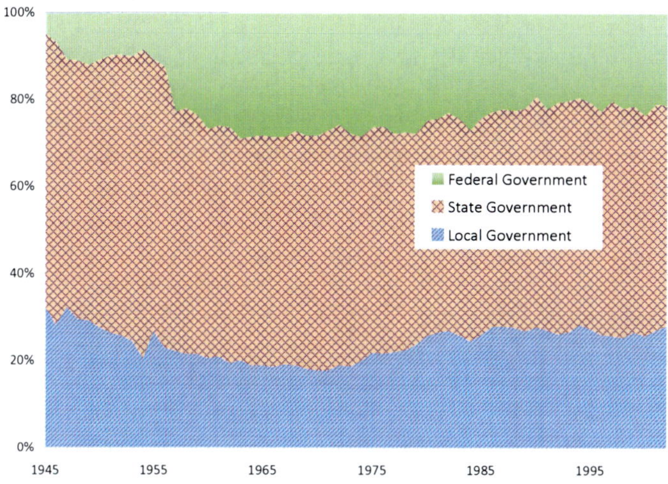

Fig. 13.1 Total receipts for highways by government levels during 1945-2002

opment. This research aims to shed some light on how regulatory efforts from different government levels, taking into account their respective pros and cons, could be reconciled in a more effective governance structure to provide an efficient, equitable, and sustainable transportation system from a long term perspective.

To address the issue of how decision power needs to be allocated at different government levels, we must first take an in-depth look at a classic question in public economics: in a federation with two levels of (local vs. central) government, which level provides public goods more efficiently? At the heart of this problem is the tradeoff between the advantages and disadvantages of alternative regimes in providing public goods. Ignoring the issues of citizens' mobility across jurisdictions,[3] local governments tend to provide public goods to a level that maximizes the aggregate surplus of their constituents, while neglecting benefits going to other districts. In the presence of positive spillovers[4] across jurisdictions, local governments will tend

---

[3] A branch of literature, dating back to Tiebout (1956), sees the advantage of decentralization as stemming from the mobility of citizens across local jurisdictions and resultant decentralized policies more closely reflecting citizens' preferences. The advantage of decentralization associated with mobility, however, is not considered in this study.

[4] Positive spillovers exist when a public good provided in one district benefits the residents of other districts. Transportation infrastructure is a public good with positive spillovers. Ignoring tolls across jurisdictions, for example, outside travelers can use local roads without paying for them, which is referred to as the "free rider" phenomenon.

to under-provide local public goods. In contrast, a central government can provide public goods more efficiently by internalizing the spillovers across local districts. The inefficiency of centralized decision-making, though, is more controversial. The classic decentralization theorem developed by Oates (1972) describes centralized provision of public goods as a process by which a central government chooses a uniform level of public good for each district, stressing that a "one size fits all" provision of public goods may fail to reflect local districts' differentiated preferences on public spending and thus undermine local interests. The modern literature, however, tends to eliminate the uniformity constraint because it is not realistic to assume a central government cannot differentiate the levels of spending in different districts. Centralized decision making may have other administrative disadvantages such as additional overhead due to increased span of control, informational asymmetries associated with local needs, etc. These factors, however, have not been formally modeled in literature.

In recent years, a rich literature of fiscal federalism has explicitly introduced political economy models to examine how conflicting local interests could result in suboptimal policy decisions in a legislature of locally elected representatives, which gives rise to another decentralizing force in the classic question of centralized versus decentralized provision of public goods. The conflicts essentially arise from distributive policy-making of a legislature that provides public goods whose benefits are geographically concentrated but costs generalized. It is expected representatives will push for high spending favored by their jurisdiction but at the expense of other jurisdictions.[5]

In political science, legislative behavior has been modeled as cooperative or non-cooperative from different perspectives, which divided the literature into two strands.[6] The cooperative view of legislative behavior assumes that representatives may exploit the benefits of universalistic cooperation in the legislature so that legislative policy decisions will maximize the benefits of all members. Nevertheless, Besley and Coate (2003) showed that suboptimal policy decisions may still result due to strategic delegation by jurisdictions in the cooperative legislature. From this standpoint, a series of legislative bargaining models has been developed to examine the inefficiency associated with strategic delegation in the cooperative legislature from different angles (Cheikbossian, 2004; Redoano and Scharf, 2004; Dur and Roelfsema, 2005; Lorz and Willmann, 2005). The cooperative views, however, neglects the difficulty in assembling and maintaining a universal coalition when members of Congress represent differing and competing demands, and when the number of representatives is large. Obviously both situations are not uncommon in multi-jurisdictional transportation systems. By contrast, a non-cooperative view of legislative behavior assumes that minimum winning coalitions of representatives

---

[5] Knight (2003), analyzing 1998 Congressional votes over transportation project funding, provided empirical evidence that legislators' probability of supporting a transportation project is increasing in own-district spending and decreasing in the tax burden associated with aggregate spending.

[6] The models of non-cooperative legislative processes originate from Buchanan and Tullock (1962); Riker (1962), while the early analytical work of cooperative legislative behavior dates back to Weingast (1979). Collie (1988) provides a survey of the two different approaches.

will form when legislative decisions are taken by majority rule. The inefficiency of legislative decision-making, as Besley and Coate (2003) pointed out, thus arises from the uncertainty associated with the identity of the minimum winning coalition, and the misallocation of resources that takes place when a non-cooperative coalition makes legislative decisions.

The above economic models of the provision of general public goods, however, are limited in two ways when applied to transportation infrastructure. First, the fiscal federalism literature in large part neglects where residents live and where they travel, therefore providing little insight into the role of spatial and demographic characteristics of geographically defined districts when investment decisions are made on a transportation infrastructure across the districts. For instance, the literature assumes a homogenous spillover effect (defined as residents' preference for public spending in other districts over the spending in their own district) for simplicity. This assumption, though, is rarely true in the case of transportation infrastructure. Indeed, individuals' preferences for spending on transportation infrastructure in other districts vary by their residential location, so do their preferences for public spending on transportation over private consumption.[7] Spatially differentiated spending preferences of individual residents may collectively play out in policy decisions through political processes. The fiscal federalism literature, however, has failed to capture the salient spatial dimension in the provision of transportation infrastructure. Furthermore, by taking residents' spending preferences as exogenous, the existing literature lacks a dynamic view on the provision of transportation infrastructure. Unlike other public goods such as schools and water treatment plants, spending on transportation infrastructure across districts changes the conditions on which current spending decisions are made (by reducing transportation cost within and across districts and thus residents' preference for transportation spending acting through the process of induced demand), which will in turn affect the spending decisions in the subsequent period. This phenomenon has been long recognized as "induced demand" by transportation professionals (Leeming, 1969; Downs, 1992).[8] Keeping in mind the mutual effects between spending decisions of local or central governments and spending preferences of individual residents, one would expect an evolving spending pattern in a sequential process of transportation infrastructure provision which the fiscal federalism literature failed to capture.

This chapter presents a spatio-economic analysis that re-addresses the question which level (local or central) of government in a two-level federation should provide transportation infrastructure, and examines the issue why a centralizing tendency

---

[7] Intuitively, one can imagine that people who drive more on roads would prefer higher public road spending, and that people who travel more outside their home jurisdiction would prefer higher road spending in neighboring districts. Further, how much people travel and travel outside their residence district depends on how far away they live from the destinations of activities located within or outside the district.

[8] Downs (1992) describes the induced demand in response to improved road infrastructure as a "triple convergence": more drivers will use an improved road because that drivers switch from other routes to the improved route (spatial convergence), that drivers who used to travel avoiding peak hours start traveling during those hours (time convergence), and that drivers who used to use public transportation start to drive (modal convergence).

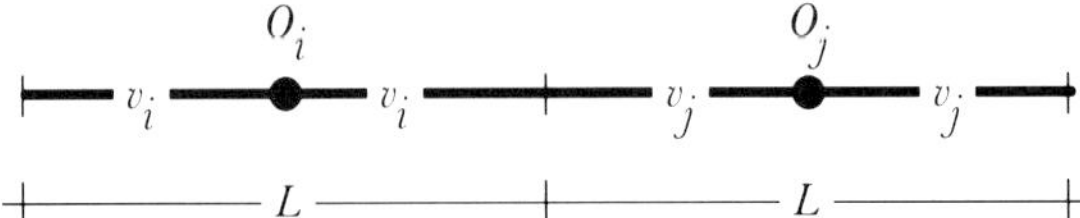

Fig. 13.2 Governance choice in two local districts

has been observed during the development of major transportation modes. While recognizing the benefits of centralization as improved coordination of interregional spillovers, this analysis introduces non-cooperative legislative decision processes in centralized provision of transportation infrastructure which may lead to inefficiency associated with uncertainty or misallocation. Local districts will weigh the benefits of alternative spending structures against their costs, and collectively decide in an autonomous political process which spending structure to adopt. This research also explores how the choice of centralized or decentralized spending structure might shift as transportation infrastructure improves and residents' preferences for spending change.

13.2 Model setting

To fix ideas, let us consider the following problem. Suppose a one-dimensional space is divided into geographically defined districts indexed by $i = 1,...,n$. A continuum of residents lives on the space. Residents travel to reach their destinations where activities (e.g., work, shopping) take place. It is assumed that activities are located only at the centers of districts. As illustrated in Figure 13.2, jurisdictions of the districts build and share a joint road infrastructure throughout the space which serve residents' travel needs within and across districts. For simplicity, it is assumed that each district maintains a uniform level of service (operationalized by road speed v_i, $i = 1,...n$) on the segment of road to its jurisdiction. Congestion is not dealt with in this analysis. Suppose the districts initially maintain the same level of service on roads (indicated by v_0). In a decentralized spending structure, local governments or districts independently choose the level of road spending. In a centralized spending structure, on the other hand, a central government takes over the decision power and determines spending levels for all the districts. While the analysis is generalizable to multiple jurisdictions, we further simplify the problem by adopting a familiar setup in the context of regional public good provision with two identical districts (indexed by i, j) to avoid redundant calculation and derive spending decisions in explicit forms. Districts have an equal size of L, represented by the distance between two boundary points of a jurisdiction. A summary of notation is given in Table 13.1.

Residents are characterized by the location of their residence x on the one-dimension space (x indicates the relative position from the location of residence to the center of the residence district). The demand function $\rho(x,i)$ represents the

Table 13.1 Notation

Variable	Description	Variable	Description
C, D	Centralization, decentralization	t, T	Travel time of individual resident, jurisdiction
d	Distance	u, U	Utility of individual resident, collective welfare of jurisdiction
E	Road construction cost	v_0	The initial road speed
f	Uniform head tax	v_i	Updated road speed chosen by district I
g, h	Derived functions	x	Distance to the center of residence district
i, j	Indices of two neighboring districts	α	Coefficients in the cost function
J	Jurisdiction	β	Weight central government puts on different districts
k	Number of trips by an individual resident in a planning period	δ	Multiplier in the demand function
L	The size of district	η	Portion of trips going to the center of residence district
M	Median voter	λ	Measure of spillover effects
n	Number of districts	ρ	Demand function
p	Planning period	τ	Value of time
r	Residential density	θ	The decay factor in the demand function

number of trips that an individual resident living at x makes to reach the activities located at the center of district i. Extending a double-cost approach,[9] the utility of an individual resident takes the following form:

At a local level,

$$u_x^D = -\tau t_x - f^D \qquad (13.1)$$

At a central level,

$$u_x^C = -\tau t_x - f^C \qquad (13.2)$$

In either case, we assume a uniform head tax will be charged on residents in order to finance road spending. Under decentralized spending we have,

---

[9] An extended double cost approach is adopted as a measure of performance in Buchanan and Tullock (1962) and Humplick and Moini-Araghi (1996*b*). It accounts for two categories of costs: the first is resource costs, which are simply the costs of provision, administration, and management of roads. In this case, we consider only the costs of road improvement; and the second is preference costs, which are defined as costs incurred by road users, in this case the monetary cost of travel time that residents have spent on roads. In line with economic theory, the objective is set to minimize the total cost, which is the sum of preference costs and resource costs.

$$\int_{-L/2}^{L/2} r_x f^D dx \equiv E_i(v_0, v_i) \tag{13.3}$$

Under centralized spending, although spending decisions are made at a central level, it is assumed road infrastructure is provided independently in different districts, so:

$$\int_{-L/2}^{L/2} r_x f^C dx \equiv 1/2 \sum_i E_i(v_0, v_i) \tag{13.4}$$

In aggregate, the welfare of residents in each district can be represented by:

$$U_i^D = -\tau T_i - E_i(v_0, v_i) \tag{13.5}$$

Where

$$T_i = \int_{-L/2}^{L/2} r_x t_x dx \tag{13.6}$$

Ideally a central government aims to maximize the aggregate welfare of residents treating districts equally. Conflicting interests of local districts, however, may play out in the legislative decision-making processes and skew road spending across districts. This issue will be addressed later.

We envisage that spending decisions on roads result from the equilibrium of a two-stage political game played by jurisdictions. In the first stage, elected representatives of districts will make the choice of policy centralization or decentralization in a representative democracy.[10] In a two-district context, a centralized structure is adopted if and only if both districts choose centralization. In the second stage, spending levels are determined at the center or in each district. If a centralized structure is adopted, a central government will take over the decision power and a legislature of locally elected representatives will determine the spending level in each district; if centralization is vetoed, each district will determine her spending level independently. Assuming a central government treats two member districts differently by putting different weights (β_i) on the two districts (how the mistreatment could occur will be explained later), the game is illustrated in Figure 13.3. As can be seen, two districts simultaneously decide which decision structure they would adopt (C represents centralization, while D represents decentralization); then spending decisions (speed levels) are made at either a central or local level.

---

[10] As Redoano and Scharf (2004) pointed out, there are two distinct political procedures for jurisdictions to determine whether or not to participate policy coordination agreements. Under a direct democracy, citizens in each region decide whether or not to centralize by referendum; under a representative democracy, on the other hand, citizens delegate the decision to elected policy makers. This analysis considers only the latter system.

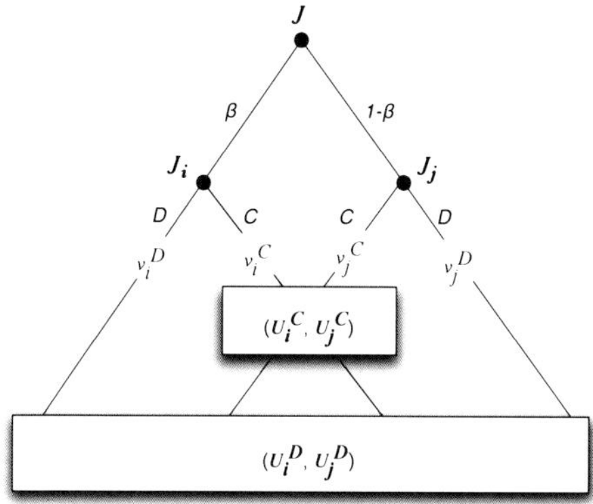

Fig. 13.3 Decision structure in a Pigouvian system of governments

Solving the game backwards, we first discuss spending decisions that will be made in a centralized versus decentralized structure. We then derive the collective choice of jurisdictions in terms of centralization versus decentralization with expected benefits and costs in alternative spending structures.

13.3 Decisions on road spending

From a normative standpoint, efficient spending decisions should maximize the total welfare of residents of interest. However, it is not realistic to expect the normative outcome will prevail when spending decisions are made politically. In this section, we start with a normative analysis which introduces a *Pigouvian* system of governments,[11] which is followed by a citizen-candidate model in which political procedures are introduced to predict spending decisions at both a local and central level.

For the sake of simplicity, it is assumed that residents are uniformly distributed across space with a density of r. It is further assumed that each individual resident generates the same number of k trips during a planning period, of which the same portion (η) of trips goes to the center of the residence district.[12] Defining the

---

[11] In political economics, there are two extreme models of local government behavior: while the Leviathan rent maximizing model assumes an extremely selfish government which seeks to maximize its own revenue, in a Pigouvian model a government represents the interests of local residents and seeks to maximize the local welfare of its population. See Epple and Nechyba (2004) for a detailed explanation.

[12] The assumption of homogenous travel demand on space, while to some extent simplifying reality, allows us to derive closed-form spending decisions analytically. Heterogeneity among residents can still be captured by their differentiated residential spatial locations and accordingly travel distances to reach activities.

spillover effect as the ratio of total travel time one spends in the neighboring district to that in the residence district, the significance of spillovers can be measured as:

$$\lambda_{x,i} = \begin{cases} 2k\eta\frac{-x}{v_0} \Big/ \Big(2k(1-\eta)(\frac{L/2-x}{v_0} + \frac{L/2}{v_0})\Big) & -L/2 \leq x \leq 0 \\ 2k\eta\frac{x}{v_0} \Big/ \Big(2k(1-\eta)(\frac{L/2-x}{v_0} + \frac{L/2}{v_0})\Big) & L/2 \leq x \leq L \end{cases}$$ (13.7)

Obviously the spillover effect varies by location, which distinguishes this spatio-economic analysis from traditional discussions in fiscal federalism in the context of general public goods.

Following Newell (1980), the demand function decays with travel cost in a negative exponential form:

$$\rho(x,i) = \delta e^{-\theta t(x,i)}$$ (13.8)

Thus the portion of trips going to the residence district can be estimated as:

$$\eta = \frac{\delta e^{-\theta \bar{t}(x,i)}}{\delta e^{-\theta \bar{t}(x,i)} + \delta e^{-\theta \bar{t}(x,j)}} = \frac{1}{1 + e^{\theta \frac{\bar{d}(x,i) - \bar{d}(x,j)}{v_0}}}$$ (13.9)

Where $\bar{d}(x,i)$ and $\bar{d}(x,j)$ represents the average distance from a resident in district i to the centers of district i and district j, respectively. Under the assumption of uniform distribution of residents over space, it is easy to check that $\bar{d}(x,i) = L/4$ and $\bar{d}(x,j) = 3L/4$. Thus,

$$\eta = \frac{1}{1 + e^{-\frac{\theta L}{2v_0}}}$$ (13.10)

The above equation defines an inverse relationship between the in-district travel demand and the initial speed of the road infrastructure. When the speed is close to zero, indicating virtual isolation between the two districts, all trips are destined for the center of the home district (η=1). When the speed increases to infinity, the two destinations are indifferent to any resident (η=0.5). With an improving road speed during transportation development, we expect to see η dropping from 1 to 0.5.

In addition, we specify a Cobb-Douglas type of cost function on road investment as follows:

$$E_i(v_0, v_i) = \alpha_0 v_0^{\alpha_1} (v_i - v_0)^{\alpha_2} L^{\alpha_3} L$$ (13.11)

Over small increments in road speed, it can be approximated as:

$$E_i(v_0, v_i) = \alpha_0 \alpha_2 v_0^{\alpha_1} (v_i - v_0) L^{1+\alpha_3}$$ (13.12)

With this, we finally have all the ingredients needed to derive the spending decisions made at a local or central level.

13.3.1 *Pigouvian governments model*

Under decentralized decision-making, benevolent local governments independently select the levels of road spending (represented by the improved road speed v_i) to maximize the total welfare of residents in the district:

$$v_i^D = \arg\max_{v_i} U_i \tag{13.13}$$

Under centralized decision-making, a central government selects the levels of road spending for each district to maximize the aggregate welfare of both districts while its preference for the districts is somehow skewed:

$$v_i^C = \arg\max_{v_i} U^C \tag{13.14}$$

If the central executive treats the benefit of each district equally, as Oates (1972) has already shown, centralized decision-making will always be superior to decentralized decision-making with the presence of positive interregional spillovers across districts and no diseconomies of scale. In reality, however, the central executive's spending decisions may be skewed as an outcome of political conflicts in legislative policy making processes.[13] Saving the discussion of legislative behavior in the next section, this model assumes a central government aiming to maximize the aggregate welfare of districts while putting different weights on them:

$$U^C = \sum_i \left(\beta_i U_i^D \right) \tag{13.15}$$

Where,

$$\sum_i \beta_i = 1 \tag{13.16}$$

Weights specified here (β_i) can be viewed as exogenous and known to districts at this stage of the game. Actually they are endogenously determined when the spending structure is chosen, which will be explained later.

Following a step-by-step procedure in Appendix A, the Nash-equilibrium road spending strategies under local and central decision making were derived as follows:

$$v_i^D = \sqrt{\frac{(1 - \eta/2)\,\tau r k L^{1-\alpha_3}}{\alpha_0 v_0^{\alpha_1} \alpha_2}} \tag{13.17}$$

$$v_i^C = \sqrt{\frac{(2 - (2 - \beta_i)\eta)\,\tau r k L^{1-\alpha_3}}{\alpha_0 v_0^{\alpha_1} \alpha_2}} \tag{13.18}$$

---

[13] For instance, Cheikbossian (2004), assuming the central government is maximizing the weighted sum of the welfare of local regions, showed that the respective weight of each locality is determined by its lobbying efforts or rent-seeking expenditures.

13.3.2 Citizen-candidate model

The Pigouvian governments model does not explicitly consider the political forces that may influence road spending decisions at a local or central level. In a representative democracy, residents in each district can vote for representatives according to their spending preferences, and delegate their spending decision to the elected representatives;[14] at a central level, a legislature constituted by locally elected representatives makes spending decisions for districts.

Under decentralized decision making, residents vote for the spending level in their residence district to maximize their individual utility. The preference of an individual resident in district i for the spending level in this district can be derived from:

$$v_{x,i}^{D} = \arg\max_{v_i} u_x^{D} \tag{13.19}$$

With majority rule and two competing political parties, it is suggested in the "median voter theorem" (Downs, 1957) that the elected representative commits to a policy position preferred by the median-type voter. In this analysis, as spending preferences of individual residents vary by their residence locations, we need to find the location(s) x_i^{M} where the median voter(s) live in order to predict the spending decision made by elected candidates.

Under centralized decision making, locally elected representatives form a legislature to determine road spending policies. From different perspectives, the legislature could be either cooperative or non-cooperative in nature. Extending Besley and Coate (2003), this analysis assumes a non-cooperative legislature in which each district elects a single representative, and each can be thought of as a minimum winning coalition while it is uncertain which coalition will be selected to determine spending policies. In the case of two identical districts, it is assumed that either representative has a chance to be selected as the policy-maker with equal probability.

When the representative from district i is selected to determine the spending policy, the representative chooses spending levels for both districts preferred by the median voter in his home district.

$$v_i^{C} = \arg\max_{v_i} u_{x,i}^{C}(x = x_i^{M}) \tag{13.20}$$

Following the procedure described in Appendix A, we found that a median voter is living in either half of a district. The distance from the location of the median voter to the center of the district was derived as:

$$\begin{cases} x_+^{M} = \frac{1}{8\eta}L \\ x_-^{M} = \frac{1-2\eta}{8\eta}L \end{cases} \tag{13.21}$$

---

[14] It is assumed in this analysis that every resident has the right to vote.

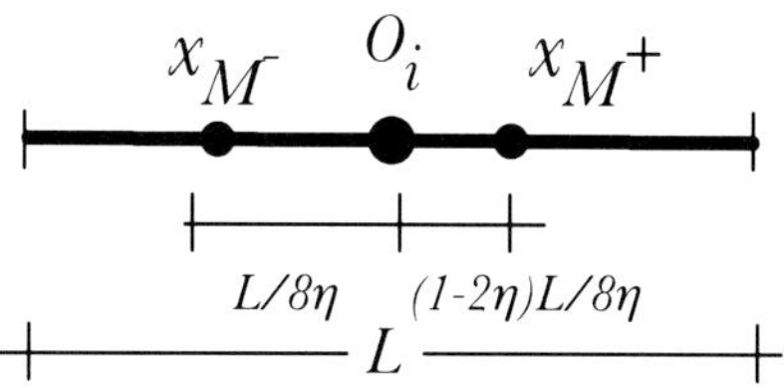

Fig. 13.4 An illustration of the endogenous choice of decision structure in a Pigouvian system of local governments.

Note that x_+^M indicates the location of the median voter living closer to the neighboring district, while x_-^M indicates the location farther from the neighboring district. Interestingly the location of the median voter (and accordingly the local spending decision) reflects the spatial distribution of travel demand. As the median voter theorem suggests, in order to become a median voter, half of the voters prefer a road spending level (surrogated by an improved road speed) that is lower than the median-type spending level. When residents travel more across the border (indicated by a smaller value of η) and demand an improvement on road infrastructure, residents living closer to the border benefit more as road spending is uniformly distributed within the jurisdiction, and thus prefer a higher spending level as compared to those living farther away from the border. Therefore, the median voter is expected to live closer to the border as η decreases. Figure 13.4 illustrates the locations of the median voters.

Accordingly, the road spending level chosen through decentralized decision making is that preferred by the median voter:

$$v_i^D = \sqrt{\frac{\left(1 - \eta + \frac{2\eta-1}{4\eta}\right)\tau r k L^{1-\alpha_3}}{\alpha_0 v_0^{\alpha_1}\alpha_2}} \tag{13.22}$$

It is easy to check that:

$$v_i^D \leq \sqrt{\frac{\left((1-\eta/2) + (1/2 - \sqrt{2}/2)\right)\tau r_i k L^{1-\alpha_3}}{\alpha_0 v_0^{\alpha_1}\alpha_2}} \tag{13.23}$$

As compared to the local spending levels determined by Pigouvian local governments shown in Equation 13.17, we can see that the election has resulted in suboptimal road spending at a local level due to competition between political parties.

Under centralized decision making, on the other hand, suppose the minimum winning coalition from district i wins the majority and takes over power in the legislature, then the representative from district i chooses levels of spending for both home and neighboring districts as:

$$v_i^C = \sqrt{\frac{2(1-\eta+\frac{2\eta-1}{4\eta})\tau rk L^{1-\alpha_3}}{\alpha_0 v_0^{\alpha_1}\alpha_2}} \qquad (13.24)$$

$$v_j^C = \sqrt{\frac{2(1-\eta)\tau rk L^{1-\alpha_3}}{\alpha_0 v_0^{\alpha_1}\alpha_2}} \qquad (13.25)$$

13.4 Choice of spending structure

With predicted spending decisions at local and central levels, representatives of each district choose the spending structure favored by their home district by comparing the expected benefits associated with centralization versus decentralization, while the final choice is made collectively. In the case of two districts, the Nash-equilibrium condition under which a centralized structure is chosen can be translated as:

$$U_i^D \leq U_i^C \text{ and } U_j^D \leq U_j^C \qquad (13.26)$$

As shown in Appendix A, if it is assumed in the Pigouvian governments model that the spending policies at a central level are skewed against district i (i.e., $0 \leq \beta_i \leq 0.5$) without loss of generality, the equilibrium condition can be derived in a reduced form that reads:

$$g(\eta,\beta_i) \leq 0 \qquad (13.27)$$

Where the specific form of $g(\eta,\beta_i)$ can be found in Appendix A.

It is worth noting that, the Ultimatum Game[15] provides insight into how the value of β_i could be endogenously determined when spending structure is chosen: the central government makes an offer (denoted by β_i, $0 \leq \beta_i \leq 0.5$) to the disfavored local government (as the favored local government will always vote for centralization) and the local government can either accept this offer (by voting for centralization) or decline it (by voting against centralization). Empirical results suggest that a fair offer (a 50/50 split) will usually be made in a Ultimatum Game,[16] while an unfair split that is less than 20% will always be declined (Henrich et al., 2004; Oosterbeek

---

[15] First studied by Güth et al. (1982), the Ultimatum Game is a bargaining game in which two players decide how to divide a sum of money: while the first player proposes a division of the money, the second player decides whether to accept it. If accepted players get their agreed upon shares; if rejected both receive nothing.

[16] The Transportation Equity Act for the 21st Century (TEA-21), for example, provided a minimum guarantee program which ensures the return of highway funding to each state is no less than 90.5% of the state's tax payment to the Highway Trust Fund (Kirk, 2004). This could be translated, in our case of two identical jurisdictions, into a Ultimatum offer of 50% with limited instability.

et al., 2004). Within this range the split may vary by the cultural and social nature of players.[17]

In the citizen-candidate model the equilibrium condition was derived as:

$$h(\eta) \leq 0 \tag{13.28}$$

The specific form of $h(\eta)$ can also be found in Appendix A.

The analysis has to this point focused on the choice of the spending structure in a static context with an exogenous road service level (represented by v_0). From an evolutionary perspective, the spending structure for a transportation system may shift over time: when the system is in its early development, decentralization is favored as transportation cost is high and spillover effects are insignificant. When the infrastructure improves and travel costs are reduced, residents travel across borders at a higher frequency, accordingly an increasing need arises to better address the issue of spillovers or free riders. When the need becomes prominent enough to overcome the inefficiency of centralization associated with legislative policy making processes, districts will eventually switch to the centralized spending structure. This idea is illustrated numerically as follows.

In the Pigouvian governments model, whether or not a centralized structure will be adopted is determined by $g(\eta, \beta_i)$ where $0.5 \leq \eta \leq 1$ and $0 \leq \beta_i \leq 0.5$. We constructed a numerical example to obtain perspective on the relationship between the choice of spending structure and the relative own-district demand (η) as well as the central government's spending preference (β_i). In doing so, we calculated $g(\eta, \beta_i)$ for each pair of η ($0.5 \leq \eta \leq 1$) and β ($0 \leq \beta_i \leq 0.5$) values with an interval of 0.01. The change of $g(\eta, \beta_i)$ is plotted against β_i and η in Figure 13.5. Note that centralization is adopted only if the governance choice function $g(\eta, \beta_i)$ has a negative value. The northeast corner of the graph, for instance, represents values of g less than 0, therefore favoring centralization. Note also that β (beta) represents the relative weight that the central government puts on the unfavorable local jurisdiction. A beta value closer to 0.5 indicates a more equitable treatment of local interests under centralization. The spillover factor η (eta) denotes the portion of trips destined for the center of home district. A smaller η indicates a more significant spillover effect, as more trips are crossing the border between districts. As can be seen in the graph, when $\beta = 0.5$, meaning the central government treats two member districts equally, a centralized spending structure is always adopted as centralized decision making outperforms decentralization by internalizing the spillovers across jurisdictions. When $\beta_i = 0$, on the other extreme, centralization is never adopted as the benefits of district i would be completely ignored by the central executive. The reality, however, likely lies somewhere between the two extremes. When $0 < \beta_i < 0.5$, as can be seen, districts may prefer decentralization when the value of η is large or the initial road speed v_0 is low (bear in mind that η has an inverse relationship with

---

[17] Our model differs from the Ultimatum Game in that the disfavored local government will get more than nothing when rejecting the offer and going for decentralization. So the central government needs to make an even higher offer (β_i) to reach an agreement than if in the Ultimatum Game.

v_0). As the road speed improves and value of η drops, the districts may switch to centralization with an increasing need for coordination. The critical value of η is larger with a higher value of β_i.

Similarly, the value of function $h(\eta)$ is calculated and plotted against η in Figure 13.6. As Figure 13.6 displays, as η decreases, the function changes from a positive value to a negative one, indicating that jurisdictions chose decentralization at the beginning, and then switched to centralization as the road infrastructure improves and travel demand across jurisdictions increases. A closer look at the numeric results discloses that the critical point of η is around 0.872.

More interestingly, as we will demonstrate later, the governance structure of a transportation system (in which the spending pattern is chosen and spending decisions made) and service conditions on infrastructure are endogenous to each other in an evolutionary process of transportation development: as infrastructure improves after spending projects are implemented, people's traveling and spending preferences change accordingly, which, in turn, will find a way into the subsequent policy decisions through a collective political system.[18] In reality, as transportation investment projects are approved and delivered in discrete planning periods, we would envision incremental improvements on infrastructure coupled with gradual changes of policy decisions in a sequential process; when a critical point is reached, a spontaneous phase-shift of the spending structure will take place, as we have observed in various transportation systems. This process is demonstrated using a numeric example of the citizen-candidate model as follows.

Suppose there are two local districts on a one-dimensional space with equal sizes (L) of 20 kilometers and equal population of 100,000. Population is uniformly distributed over the space with a residential density (r) of 5,000 persons per kilometer. We assume an average person generates 10,000 trips (k) from home in one planning period (say 10 years) and has a value of time (τ) of \$10 per hour. A road crosses both districts with an initial speed (v_0) of 35 kilometers per hour. While fixing the travel demand (k), we assume the population in both districts is growing exogenously at a constant rate of 0.15 for each planning period. We arbitrarily specify the decline factor θ in the own-district demand function as 10.0, and the values of the cost function coefficients α_i $(i = 0,1,2,3)$ as 1,000, 1.25, 1.25, and 0.95, respectively. It needs to be noted that while the specified values of the parameters capture some aspects of reality, they have not been calibrated with empirical data. They are only used to demonstrate the idea.

Simulation results are displayed in Table 13.2. For each planning period p the split of own-district demand (η) is estimated according to the current population $(r(p))$ and road speed $(v(p))$. The function of $h(\eta)$ is then computed, and a centralized or decentralized spending structure is adopted according to its score. Finally, policy decisions are made in the chosen structure which, upon implementation, update the road speed for the next planning period $(v(p + 1))$. As can be seen, in the beginning rounds the road service level (speed) is low and travel is mostly intra-

---

[18] Besides infrastructure conditions, many factors such as population growth, technological advance, and induced demand may influence spending decisions. Including these factors endogenously in our models, however, is beyond the scope of this analysis

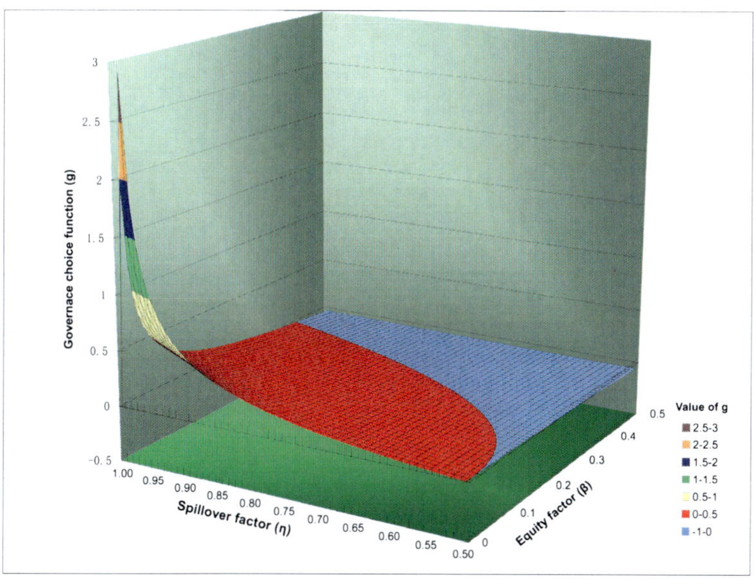

Fig. 13.5 A numeric illustration of the endogenous choice of decision structure with the change of η and β_i in the Pigouvian governments model. A negative value of $g(\eta, \beta_i)$ indicates the adoption of a centralized structure.

district (with η close to 1.0); so a decentralized spending structure is always chosen under which local governments make incremental improvements on their own road segments. This lasts until the seventh consecutive planning period, when η reaches the critical value of 0.872, and the spending power is shifted to the central government. Uncertainty subsequently arises regarding the updated road speeds under centralized control, depending on who will win the majority and take over decision power in the legislature.

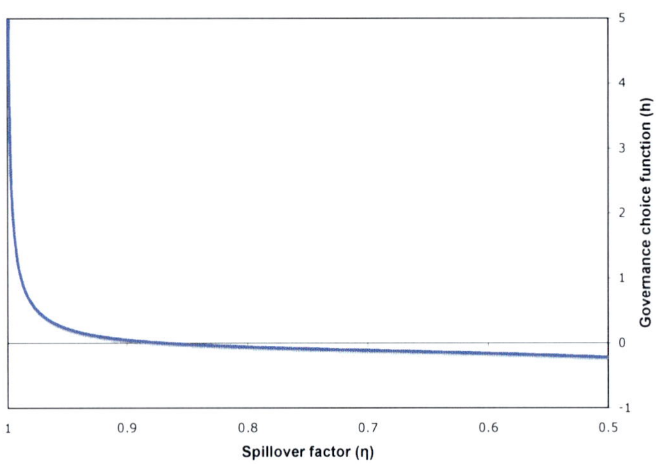

Fig. 13.6 A numeric illustration of the endogenous choice of decision structure with the change of η in the citizen-candidate model. A negative value of $h(\eta)$ indicates policy centralization.

Table 13.2 A numeric example of the coupled evolution of infrastructure and the governance structure in a sequential process

p	r(p)	v(p)	η	h(η)	Centralized	v(p+1)
0	5000	35.000	0.946	0.191	No	39.781
1	5750	39.781	0.925	0.106	No	40.366
2	6613	40.366	0.923	0.098	No	43.022
3	7604	43.022	0.911	0.067	No	44.923
4	8745	44.923	0.903	0.049	No	47.319
5	10057	47.319	0.892	0.030	No	49.668
6	11565	49.668	0.882	0.014	No	52.210
7	13300	52.210	0.872	-0.0005	Yes	N.A.

13.5 Findings and concluding remarks

While transportation economists, public economists, and political scientists have examined governance choice in the provision of public goods from different perspectives, this study constructed a holistic framework that accounts for the spatial, economic, and political dimensions of transportation infrastructure provision. This framework could be extended to examine the efficiency of alternative financing instruments (such as tax versus toll) in addressing contemporary transportation issues (congestion pricing, capacity expansion with limited resources, etc.) with governmental involvement.

In this framework, we analyzed governance choice in a two-level federation in the provision of road infrastructure across jurisdictions. Both governmental and individual decisions are examined. It is demonstrated, on a serial road network shared by two districts, that residents' preferences for public spending on road infrastructure are differentiated by where they live and where they travel. In consideration of differentiated travel demand and spending preferences over space, a two-stage imperfect information game is then constructed to predict the choice of centralized or decentralized spending structure in a representative democracy, and the corresponding spending decisions at either a central or a local level.

Two models have been proposed. While the first model considers simple Pigouvian behavior of governments, the second explicitly models political forces at both government levels.

At a local level, residents in each district vote for representatives and delegate their spending decision to elected representatives. Under the assumption of two-party politics and majority rule, elected representatives will commit to a policy position preferred by the median voter. It is found that the location and preferred spending level of the median voter depends on in-district travel demand; it is also found that elected representatives, in order to win the majority in the election, tend to provide a lower level of spending as compared to Pigouvian local governments.

At a central level, it is assumed that spending decisions are made in a legislature of locally elected representatives, where tradeoffs have to be made between the uncertainty and misallocation associated with non-cooperative legislative policy making, and its advantage in terms of better addressing spillovers across districts. Both models led to the finding that a centralized or decentralized spending structure is chosen based on a satisfactory compromise between benefits and costs associated with alternative decision making processes, and that the governance structure may shift as the infrastructure improves over time.

The goal of this study is to capture the essence of governance choice in a spatio-economic context. In order to derive spending decisions in explicit forms, we simplified reality by assuming a homogenous travel demand over space and a non-cooperative legislature in this analysis. This could be extended by introducing more realistic travel demand models and legislative behaviors. In addition, although it has focused on the symmetric case with two identical districts, this analysis is generalizable to examine multiple jurisdictions in an asymmetric situation using numeric evaluation, for which results may differ significantly.

The implications of this study are several. This research considers individual transportation decisions based on the costs generated and the benefits received from their mobility actions, and examine governance choice and road spending decisions as a function of individuals' preference in a democratic referendum system. From an evolutionary perspective, this analysis points out that the governance structure is not static. As transportation infrastructure improves after rounds of investments and technological advance, the desired choice may spontaneously switch from decentralization to centralization. This explains the observation that the development of major transportation modes in the US was usually accompanied with a centralizing tendency of decision power, and provides perspective on the coupled evolution of transportation infrastructure and governance choice arising from supply-demand interactions. Last but not least, this study contributes to a better understanding of the benefits and costs associated with governance at different government levels, and how their regulatory efforts could be reconciled toward the goal of sustainable transportation development.

Chapter 14
Governance Choice - A Simulation Model

14.1 Introduction

An electrical engineer in training, Paul Baran joined the RAND Corporation in 1959. While working at RAND on a scheme for U.S. telecommunications infrastructure to survive a "first strike," he conceived of the Internet and digital packet switching, the Internet's underlying data communications technology. Using minicomputer technology of the day, Baran (1964) demonstrated in a simulation suite that a distributed packet switched communications network can better withstand massive destruction to individual components than centralized or decentralized networks. Baran's conceptual drawings in Figure 14.1 serve as a vivid example of the distinctly different network patterns under centralized versus decentralized control.

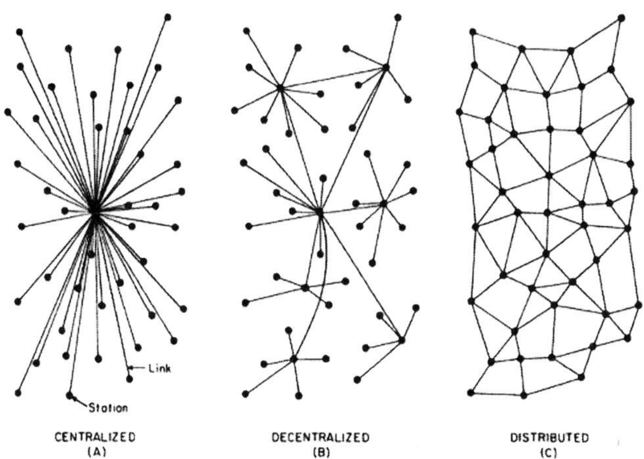

Fig. 14.1 Baran's conceptual networks for data communications (Baran, 1964)

This chapter explores transportation networks under centralized versus decentralized control. Focusing on the governmental provision of transportation infrastructure, Chapter 13 presented a game-theoretic analysis of governance choice at a central versus local level, considering the benefits and costs associated with alternative governance patterns. In order for mathematical tractability, however, this analysis only examined an idealized, small serial network. Chapter 10 constructed System Of Network Incremental Connection (SONIC) which created a transportation network link-by-link during its early deployment phase, although this model, growing a network based on the myopic "strongest-link" heuristic, does not consider the institutional organization that underlies this process.

Extending the efforts in the previous chapters, this chapter aims to construct a more realistic representation of governmental control on transportation networks. Adopting the basic form of the SONIC model, this chapter develops System Of Network Incremental Connection for Governance Choice (SONIC/GC), which predicts the incremental deployment of a transportation network under alternative governmental initiatives, and compares the resulting system performances. While it is generalizable to any surface transportation network, the model is developed with a particular focus on road networks.

The complexity in modeling the provision of transportation networks arises from the many actors who design, construct, expand, manage, maintain, operate, commercialize, and use transportation networks during the evolutionary process of transportation development. Stripped to its essence, transportation infrastructure is provided and operated in a value chain in which three key groups of players, including customers (travelers), financiers (bank), and providers (in this case, central or local governments), are involved. Therefore, the deployment of a transportation network is played out as the outcome of the strategic decisions made by these players under their independent initiatives. Main assumptions regarding the players are laid out below.

Travelers prefer a route on a transportation network that incurs less generalized travel cost, which includes travel time and monetary costs they pay for travel, such as parking, fuel taxes and user tolls. A deterministic behavioral mechanism assumes travelers have perfect information regarding travel time over the entire network and they always choose the least cost route from their origin to destination. A stochastic theory, on the other hand, relaxes this assumption and includes a random component in travelers' perception of travel time, assuming that travelers choose routes to minimize their perceived travel cost.

A *bank* allows jurisdictions to save the surplus of their revenue for future investment, or borrow from the future for present spending. The bank pays interest for the savings and provides loans at an interest rate. It is assumed that a rational bank would prioritize its loans to investment projects with funding needs trading off risk for reward. For simplicity, this study assumes a central bank agent and no spread between the rate for savings and the rate for lending.

Providers of transportation infrastructure could be public or private, at a central or a local level. This analysis concentrates on its public provision led by central or local governments, which was the common practice for road financing during the twentieth century in the United States. Although a mixed governance structure is common in reality, this study treats centralized and decentralized provision separately. It is assumed a central or local government as a perfect Pigouvian provides transportation goods to maximize the aggregate welfare of its constituents (Levinson, 2002). Central to a realistic representation of governmental control, SONIC/GC implements the key ingredients of pricing and investment policy-making as follows:

- *Pricing*: Central or local governments set tolls or taxes for the use of transportation facilities to their jurisdictions, and may adjust the rates in response to the variations to the demand. Travelers, on the other hand, may respond to the added monetary costs (tolls or taxes) associated with their travel by adjusting their destination or route choices. Although tolls or taxes may be levied for the use of infrastructure as a means of financing, a Pigouvian government does not intend to maximize its toll revenue, because user tolls or taxes increase access cost to properties, therefore reducing windfall gains in land values for infrastructure provision, and the end effect on overall welfare would be the same (Mohring and Harwitz, 1962).

- *Investment*: Governments invest in both expansion on existing roads and construction of new roads. To make it more realistic, the model allows governments to choose not only the location but also the amount (capacity) of road investment on a network in discrete time periods. It is assumed governments prioritize investment projects on transportation networks according to their cost-effectiveness, and build the most cost-effective ones until the funds (which come from toll revenue and last period's surplus, if any) are exhausted. Although an intelligent government should be able to rank expansion projects against new construction projects in terms of their funding priorities, this study separates expansion budget from a given portion of annual budget for the purpose of simplicity. Under decentralized control, a cross-border infrastructure project may involve multiple local jurisdictions, which necessitates a negotiation regarding how the cost and revenue associated with the project will be split. Once a project is built, whether it is funded by a single jurisdiction or by the joint venture of several, it is assumed that a project operator who represents the owner(s) will manage and maintain this project as a whole.

Given the complexity it involves, this research is not intended to be comprehensive. Instead, it focuses on providing a modeling tool which represents the deployment of a road network under governmental control, and demonstrates the capability of assessing alternative pricing and regulatory regimes from a normative perspective.

The remainder of this chapter is organized as follows: the next section will describe the model in details. Numeric experiments will then be conducted on an idealized network, which is followed by a discussion of the results. The conclusion section summarizes findings and suggest directions for future research.

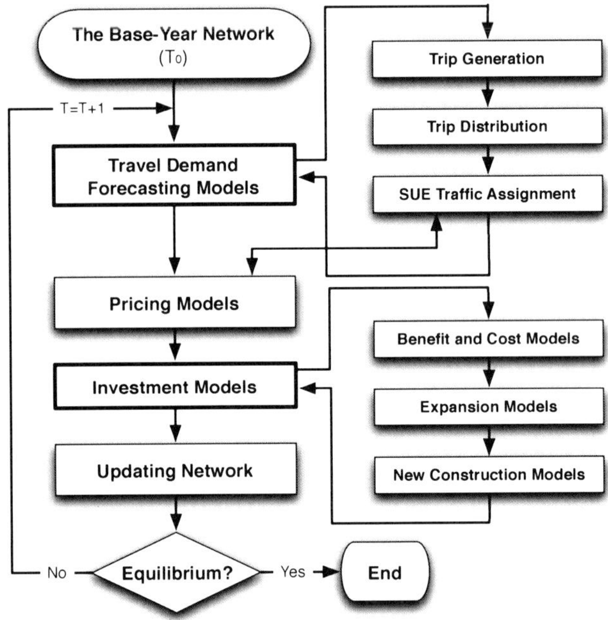

Fig. 14.2 Model framework of SONIC/GC: System Of Network Incremental Connection for Governance Choice

14.2 Model

System Of Network Incremental Connection for Governance Choice (SONIC/GC) is constructed to represent the incremental deployment of a transportation network under alternative governmental control. As illustrated in the flowchart of Figure 14.2, travel demand models predict daily traffic across individual links on a given network; pricing models set and adjust toll rates on roads also on a daily basis while toll revenue accrues to central or local governments annually; based on benefit and cost estimates, expansion models improve existing congested roads, while new construction models construct new roads subject to budgetary constraints. An integration of these component models predict the daily choice of destinations and routes of individual travelers on a road network, as well as the policy decisions of central or local governments with regard to the financing, investment, maintenance and operation of the network in the long run. The process is iterated until neither expansion nor new construction occurs or up to a maximum of 50 iterations.

Variables and coefficients that will appear in the remainder of this chapter are summarized in Tables 14.1 and 14.2, respectively.

Table 14.1 Notation: Variables

Variable	Discription	Variable	Discription
A_i	Accessibility to jobs of place i	m,n	Index of subgraph
a	Index of link	N	Planning horizon in year
B	Net social benefit	O_i	Trip generation of Zone i
C_a	Link capacity	P	Set of places
c,d	Ownership (centralized, decentralized)	p,q	Index of road project
D_i	Trip attraction of place i	Q^*	Threshold volume capacity ratio
E_e	Construction cost, expansion projects	R_p	Collected revenue of road project p
E_n	Construction cost, new projects	r_s,r_l	interest rates for saving and loan
f_a	The flow of link a	s_a	Free flow speed of link a
G	The complete graph	T_{ij}	Number of trips from place i to place j
G_m	Subgraph	$t,\tilde{t}$	Actual travel time, generalized travel cost
H	Benefit-cost ratio	t_0	Intra-place travel time
i,j	Index of place	U_i	Land use of place i
J_i	Number of jobs in place i	V	Set of vertices
K	Balancing variables in trip distribution	v	Index of vertices
k	Index of iteration	W_i	Number of resident workers in place i
l_a	The length of link a	Y_a	Travelers' willingness to pay on link a
L	Set of links	Z	Current balance

14.2.1 Travel demand models

Consistent with the SONIC model in Chapter 10, travel demand models include trip generation, trip distribution, and traffic assignment while skipping mode choice by assuming a single mode of vehicular trips. It is assumed land use activities are distributed at a given set of places. Trips originating in and attracted to each place are positively correlated to the intensity of land use activities. The analysis herein considers only two types of land use activities: residences and businesses. Thus trip production or attraction is simply estimated in a linear equation as follows:

$$O_i = \xi_0 + \xi_1 J_i + \xi_2 W_i \tag{14.1}$$

$$D_i = \psi_0 + \psi_1 J_i + \psi_2 W_i \tag{14.2}$$

A doubly constrained trip distribution procedure is implemented to predict the travel demand between a pair of origin and destination places. The interaction between places assumes a gravity-type negative exponential form:

Table 14.2 Notation: Parameters

Parameter Description

ξ_0, ξ_1, ξ_2	Coefficients in trip generation
ψ_0, ψ_1, ψ_2	Coefficients in trip attraction
ω_1, ω_2	Coefficients in empirical capacity-speed relation
υ	Value of time
θ	Spatial interaction coefficient
μ	Expansion share of investment budget
κ	MSA convergence threshold
ρ_1, ρ_2	Toll rate adjustments under decentralized control
$\beta_0, \beta_1, \beta_2$	Coefficients in empirical construction cost models
ϕ	Coefficient that coverts peak hour traffic to daily traffic
χ_p^i	Ownership share of place i in project p
δ_a^p	Dummy variable denoting if link a belongs to project p
δ_a^{ij}	Dummy denoting if link a is on the least-cost path from place i to j
π	Value of accessibility to jobs
λ_a	Proportion of resident-travelers on link a
ε	Constant

$$N_{ij} = K_i K_j O_i D_j e^{-\theta \tilde{t}_{ij}} \tag{14.3}$$

The calculation of the generalized travel cost $\tilde{t}_{ij}$ adopts the same form as in Equation 10.7.

Traffic assignment adopts the basic procedure of Stochastic User Equilibrium (SUE) (Sheffi, 1985), while also including variable tolls. Supposing, in an extremely decentralized pattern of road pricing, individual road operators set tolls on their subordinate links independently and adjust toll rates solely according to through traffic ($\tau_a = \tau_a(f_a)$), equilibrium is reached when neither travelers nor road operators would deviate their decisions unilaterally. A revised Method of Successive Average (MSA) procedure is then proposed as follows to pursue this equilibrium:

- Step 0: Perform a stochastic network loading procedure based on $\{\tilde{t}_a^0 = t_a^0 + \tau_a^0/\upsilon\}$, the set of initial generalized travel times from the resulting generalized travel times of the preceding time period, which generate a set of link flows $\{f_a^1\}$. Set $n:=1$
- Step 1: Update toll on each link $\tau_a^n = \tau_a(f_a^n), \forall a$.
- Step 2: Update link travel time on each link $t_a^n = t_a(f_a^n), \forall a$.
- Step 3: Perform a stochastic network loading procedure based on the current set of generalized link travel times $\{\tilde{t}_a^n = t_a^n + \tau_a^n/\upsilon\}$, which generates an auxiliary link flow pattern $\{\hat{f}_a^n\}$.
- Step 4: $f_a^{n+1} = f_a^n + (1/n)(\hat{f}_a^n - f_a^n)$
- Step 5: Stop upon convergence or set $n:=n+1$ and go to Step 1.

This study sets the convergence rule with a maximal allowable link flow change between two consecutive network loadings and implements it within a maximum of 150 iterations:

$$\left| f_a^{n+1} - f_a^n \right| < \kappa, \forall a \tag{14.4}$$

Powell and Sheffi (1982) have proven that the convergence of MSA for SUE is ensured only if $\tilde{t}_a(f_a)$ and $d\tilde{t}_a(f_a)/df_a$ are strictly positive and bounded for feasible values of f_a. Without a toll the convergence conditions are commonly met in practice, for instance, with the fourth power U.S. BPR curve (Bureau of Public Roads, 1964):

$$t_a = (l_a/s_a)(1 + 0.15(f_a/C_a)^{4.0}) \tag{14.5}$$

Road pricing will be discussed in the next section in more details.

14.2.2 Pricing models

Charging for the use of public roads provides road suppliers a source of income as well as a means of recovering road investment. Suppose that the travel demand on a link depends solely on its flow, that the inverse demand curve is indicated by $Y_a(f)$, and that the proportion of trips made by local residents at a given point of time is λ_a (in other words, the volume of trips made by residents is $\lambda_a f_a$ while that by non-residents is $(1 - \lambda_a)f_a$). It is further assumed that λ_a is fixed over a small change of flow or toll rate. For the jurisdiction that operates this link, its net social benefit from the link can be written as:

$$B = \lambda_a \int_0^{f_a} (Y_a(f) - t_a - \tau_a)df + f_a \tau_a \tag{14.6}$$

Note that while the toll is imposed without discriminating between resident and non-resident travelers, the toll revenue from residents is viewed as a transfer within the jurisdiction, and thus cancels out with residents' toll payment in the net social benefit equation. To maximize the benefit, set the first derivative of Equation 14.6 with regard to λ_a at zero which then yields:

$$Y_a(f_a) = t_a(f_a) + \frac{\partial t_a(f_a)}{\partial f_a}f_a + \frac{\lambda_a - 1}{\lambda_a}\left(\tau_a(f_a) + \frac{\partial \tau_a(f_a)}{\partial f_a}f_a\right) \tag{14.7}$$

The left-hand side of the equation represents the marginal benefit of producing an extra trip, while the right-hand side represents the marginal cost. The marginal cost is the summation of three components which, from left to right, represent average travel cost per trip, the change in average travel cost from serving an additional trip, and the change in toll rate, respectively. In order to maximize the net social benefit, a toll needs to be set such that:

$$\tau_a(f_a) = \frac{\partial t_a(f_a)}{\partial f_a} f_a - \frac{1-\lambda_a}{\lambda_a}\left(\tau_a(f_a) + \frac{\partial \tau_a(f_a)}{\partial f_a} f_a\right) \tag{14.8}$$

In another form:

$$\tau_a(f_a) + (1-\lambda_a)\frac{\partial \tau_a(f_a)}{\partial f_a} f_a - \lambda_a \frac{\partial t_a(f_a)}{\partial f_a} f_a = 0 \tag{14.9}$$

This is a partial differential equation. If $t_a(f_a)$ is specified as a BPR function as in Equation 14.5, the solution to this equation is then given by:

$$\tau_a(f_a) = \frac{0.6\lambda_a}{5-4\lambda_a}(l_a/s_a)(f_a/C_a)^{4.0} + \varepsilon f_a^{-\frac{1}{1-\lambda_a}} \tag{14.10}$$

Unfortunately, the solution is not unique due to the unspecified constant ε.

When $\lambda_a = 1$, meaning that all the travelers are local, the toll rate is set at the marginal travel cost as follows:

$$\begin{aligned}
\tau_a(f_a) \\
= \frac{\partial t_a(f_a)}{\partial f_a} f_a \\
= \frac{\partial\left((l_a/s_a)(1+0.15(f_a/C_a)^{4.0})\right)}{\partial f_a} f_a \\
= 0.6(l_a/s_a)(f_a/C_a)^{4.0}
\end{aligned} \tag{14.11}$$

This applies to centralized control as the central government treats every resident as local. Although fuel tax is still the most common practice throughout the United States, marginal cost pricing has been the subject of academic interest for decades as the first-best optimal pricing strategy in theory (Mohring and Harwitz, 1962; Gómez-Ibáñez et al., 1999), and started to gain popularity among practitioners in recent years. The above equation represents the marginal-cost pricing function that has been derived in the one-link static scenario under centralized jurisdictional control. Note that the equation is derived based on the assumption that the free flow speed and capacity are fixed in the short run, and the toll rate adjusted solely depending on through traffic.

Anderson and Mohring (1997) computed marginal congestion costs on the road network of the Minneapolis-St. Paul metropolitan area using a link-by-link method, and proposed a marginal congestion pricing policy based on the assumption that marginal congestion costs for each link could be used as substitutes for the true system-wide marginal congestion costs. Following Anderson and Mohring (1997), this study adopts a link-by-link marginal-cost pricing policy under centralized jurisdictional control, by which the central government sets the marginal-cost price on individual roads as described in Equation 14.11 as if each road were operated in the one-link static environment.[1] In this case, the generalized travel time on a link can be written as:

$$\tilde{t}_a = t_a + \tau_a^c / v$$
$$= (l_a/s_a)(1 + (0.15 + 0.6v^{-1})(f_a/C_a)^{4.0})$$
$$(14.12)$$

According to Powell and Sheffi (1982), the necessary conditions for the convergence of MSA in traffic assignment are satisfied.

When $\lambda_a = 0$, the solution becomes:

$$\tau_a(f_a) = -\frac{\partial \tau_a(f_a)}{\partial f_a} f_a \qquad (14.13)$$

This is the toll rate that maximizes toll revenue ($\tau_a f_a$) when all the travelers are non-residents, which is the case when the road operator is private. Note that it is assumed $\frac{\partial \tau_a(f_a)}{\partial f_a} < 0$, indicating that the toll rate decreases as the volume of trips increases.

When $0 < \lambda_a < 1$, a road is traversed by a mix of resident and non-resident travelers. This applies when roads are provided by local jurisdictions. In such a decentralized pattern, it is assumed that individual links are operated under the interests of local jurisdictions, and that local road operators have to set and adjust toll rates independently without knowing others' decisions. Due to the imperfect and incomplete information involved in this process, a heuristic price-probing method is proposed to predict toll setting decisions on individual roads.[2] The implementation procedure of local operators' toll-setting behaviors is presented as follows, and embedded in the toll-updating step of the aforementioned revised MSA algorithm to pursue demand-performance equilibrium:

- Step 0: The initial toll rate is estimated using marginal-cost price with the flow adopted from the preceding time period (use estimated flow for a new link).

- Step 1: In the second MSA iteration, each road operator attempts to increase its toll rate by ρ_1, as the operator knows the price should be somewhere between the marginal cost price and the higher profit-maximizing price but doesn't know what the exact increase should be.

---

[1] It should be noted that, however, the complexity of marginal-cost pricing goes beyond the one-link static model adopted in this analysis. Considering network effects, the demand on a link depends not only on the cost of traversing this particular link, but also on the costs of using other links as its cooperators or competitors, which becomes too complex to be specified as an equation. Safirova et al. (2007), comparing marginal congestion costs computed link-by-link with measures taking into account network effects, found that while network effects are not significant in the aggregate, marginal cost measured on a single link does not accurately predict the actual congestion cost on that link.

[2] Bayesian Nash equilibrium may be derived analytically where only a small set of links are involved in toll setting, but the game theoretic problem becomes almost unsolvable for a real-size network.

- Step 2: In iteration n, the toll rate is updated based on the information derived from previous iterations:

  - Step 2.1: the proportion of resident-travelers on each link (λ_a^n) is estimated and updated after each network loading. Since it is assumed that the operator of a link represents a joint venture of the jurisdictions that own this link, the number of travelers from a jurisdiction is weighted according to the jurisdiction's respective ownership share on the link and summed up to the total number of resident travelers using this link.
  - Step 2.2: the change of net social benefit (ΔB_a^{n-1}) from iteration $n-2$ to $n-1$ is estimated. As the travel demand changes in response to the increase or decrease in the generalized travel time, with relatively small changes in cost, the change of consumer surplus can be approximated using the "rule of 1/2" (Neuberger, 1971; Xie and Levinson, 2007b) as:

$$\Delta CS_a^{n-1} = 0.5(\tilde{t}_a^{n-2} - \tilde{t}_a^{n-1})(f_a^{n-1} + f_a^{n-2}) \tag{14.14}$$

Since the operator is concerned with both consumers' surplus from resident-travelers and toll revenue from non-residents, the change in total social benefit can be estimated as:

$$\Delta B_a^{n-1} = 0.5(\tilde{t}_a^{n-1} - \tilde{t}_a^{n-2})(\lambda_a^{n-1} f_a^{n-1} + \lambda_a^{n-2} f_a^{n-2}) + f_a^{n-1} \tau_a^{n-1} - f_a^{n-2} \tau_a^{n-2} \tag{14.15}$$

  - Step 2.3: a link changes its toll rate by $\Delta \tau_a^n = \tau_a^n - \tau_a^{n-1}$ according to the following myopic rules:

    If $\Delta B_a^{n-1} = 0$, the operator would expect its benefit may be at a local maximum, so the toll rate remains unchanged ($\Delta \tau_a^n = 0$);

    If $\Delta B_a^{n-1} > 0$, then $\Delta \tau_a^n > 0$, meaning if the net social benefit increased during the last iteration, the operator would keep the direction of toll adjustment. Suppose in this case it will adopt a conservative pricing policy which increases its toll at a decreasing rate ($0 < \rho_2 < 1$) in order to approach a local maximum:

$$\Delta \tau_a^n = \rho_2 \Delta \tau_a^{n-1} \tag{14.16}$$

If $\Delta B_a^{n-1} < 0$, then $\Delta \tau_a^n < 0$, meaning if the net social benefit decreased, the operator would change the direction of toll adjustment. In this case, if $\Delta B_a^{n-2} > 0$, the operator would expect a local maxima that lies somewhere in between, so a toll could be set as:

$$\Delta \tau_a^n = -\Delta \tau_a^{n-1} |\Delta B_a^{n-1}| / (|\Delta B_a^{n-1}| + |\Delta B_a^{n-2}|) \tag{14.17}$$

Otherwise, if $\Delta B_a^{n-2} < 0$, the benefit has decreased since iteration n-2, so the operator would adjust the toll further back beyond τ_a^{n-2}.

$$\Delta \tau_a^n = -\Delta \tau_a^{n-1} (|\Delta B_a^{n-1}| + |\Delta B_a^{n-2}|) / |\Delta B_a^{n-1}| \tag{14.18}$$

14.2.3 Investment models

The implementation of road investment is illustrated in Figure 14.3. Under either centralized or decentralized road provision, it is assumed investment decisions are made based on a benefit-cost analysis. For each time period, road investment projects with the highest benefit-cost ratios will be built first until the budget (estimated from toll revenue) is exhausted. It is assumed a given portion (μ) of annual income is separated for expansion and the remaining for new construction. The component models of revenue, benefit, cost, expansion, and new construction are explained in turn as follows:

14.2.3.1 Toll revenue

Under centralized control, the central government collects toll revenue from all the public roads. Suppose roads are built and managed in projects, the annual income of a central government can be calculated as:

$$R^c = \sum_p R^c_p = \sum_p \sum_a (365 \phi \delta_a^p \tau_a^c f_a) \tag{14.19}$$

Under decentralized control, on the other hand, a road operator collects toll revenue from its subordinate links and the remaining revenue after necessary road spending eventually accrues to the balance of the owner(s).

$$R^d_p = \sum_a (365 \phi \delta_a^p \tau_a^d f_a) \tag{14.20}$$

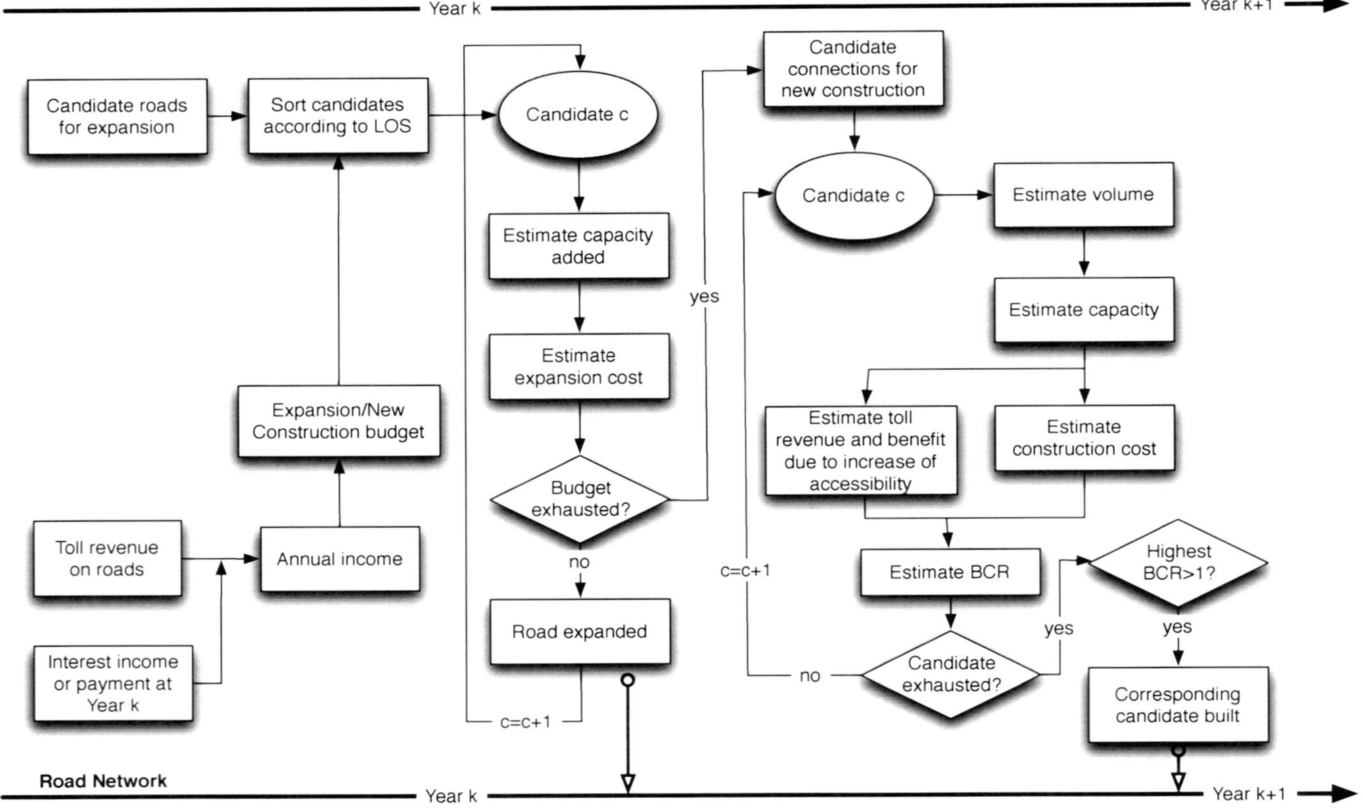

Fig. 14.3 Implementation of road investment process

14.2.3.2 Benefit

In the centralized case, it is assumed that benefits associated with road investment comes from the increased property value throughout the region, as transportation investment leads to the reduction of access cost to properties, and to increased attraction and accrued land value in the long term. The central government uses toll only as a means of financing transportation expenditures. By contrast, local jurisdictions charge user tolls on the infrastructure as otherwise outside travelers would use it for free, so under decentralized control, the benefits of road investment come both from the increased value of local properties and from projected toll revenue from non-residents. Tolls from a jurisdiction's local residents are simply considered transfers, and are dropped from the benefit calculations for both centralized and decentralized cases.

For the purpose of this study, it is assumed accessibility to jobs surrogates the long-term desirability of a place by calculating job opportunities that can be reached from this place via a road network impeded by the travel cost on the network. The accessibility to jobs is computed using a gravity-type measure in the following form:

$$A_i = W_i \sum_j J_j e^{-\theta \tilde{t}_{ij}} \tag{14.21}$$

It is further assumed that the land value of a place is estimated by pricing accessibility at a constant rate according to:

$$U_i = \pi A_i \tag{14.22}$$

The monetary value of a unit of gravity-type spatial accessibility to jobs is estimated in a recent empirical study by El-Geneidy and Levinson (2006) based on 44,429 home sale records for the year of 2004 in the Twin Cities metropolitan region. A hedonic model discloses the relation between single-family residence property values and accessibilities to jobs and to residents with other factors controlled. In essence the capitalized value of access in home prices reflects the value of travel time saved in the long run.

14.2.3.3 Cost

Cost functions estimate the monetary cost of proposed expansion or new construction (spending on maintenance is neglected for simplicity) on roads. The expansion and new construction models below adopt empirical cost functions estimated by Levinson and Karamalaputi (2003*a*) while assuming that any project can be completed within one year.

$$ln(E_{e,p}) = \beta_0 + \beta_1 ln(\sum_a \delta_a^p l_a \Delta C_a) \tag{14.23}$$

$$ln(E_{n,p}) = \beta_1 ln(\sum_a \delta_a^p l_a \Delta C_a) + \beta_2 \tag{14.24}$$

Note that the costs above are measured in thousands of dollars. After the costs of expansion and new construction are appropriated, remaining annual income accrues to a jurisdiction's current balance. Under centralized control, the balance of the central government at the beginning of iteration $k+1$ is calculated as follows:

$$Z^{c,k+1} = Z^{c,k}(1+r) + R^{c,k} - E_e^k - E_n^k$$
$$Where: \; r = \begin{cases} r_s, \; if \, Z^{c,k} > 0 \\ r_l, \; if \, Z^{c,k} < 0 \end{cases} \tag{14.25}$$

In the decentralized scenario, the toll revenue and expansion cost of a project are split between owner jurisdictions according to their respective shares (how the shares could be determined will be discussed later):

$$Z_i^{d,k+1} = Z_i^{d,k}(1+r) + \sum_p (\chi_p^i(365\phi R_p^{d,k} - E_{e,p}^k)) - \sum_q (\chi_q^i E_{n,q}^k)$$
$$Where: \; r = \begin{cases} r_s, \; if \, Z_i^{d,k} > 0 \\ r_l, \; if \, Z_i^{d,k} < 0 \end{cases} \tag{14.26}$$

14.2.3.4 Expansion

The central government under centralized control or local governments under decentralized control select projects for expansion and determine the amount of capacity addition following the procedure described below:

- Step 0: Congested links with their volume-capacity ratios above a threshold (Q^*) constitute a set of candidates for expansion.
- Step 1: Sort candidate links based on volume-capacity ratios from high to low.
- Step 2: Expand the first link among remaining candidates by the amount of :

$$\Delta C_a = f_a/Q^* - C_a \tag{14.27}$$

.
- Step 3: Deduct the expansion cost ($E_e(\Delta C_a)$) from the budget for road expansion.
- Step 4: If expansion budget has not been exhausted and there are remaining candidate projects, go to Step 2; otherwise stop.

14.2.3.5 New construction

Central to incremental network growth is the addition of new links in a sequential process, which is implemented as follows:

First, find the exhaustive set of candidate projects for new construction including both internal and external connections. A potential connection between a pair of established nodes is identified as a candidate project if and only if no established links exist along the geographical shortest path between the two nodes. To illustrate, three candidate projects are identified in Figure 10.2 corresponding to connections $v_3 - v_4$, $v_5 - v_7$, and $v_8 - v_9$. Although not fully connected by established links, the connection $v_4 - v_5 - v_7$ is not identified as a candidate project because it contains an established link $v_4 - v_5$ (instead, $v_5 - v_7$ is identified as a candidate). To reduce the size of the candidate set, intuitive rules are proposed: it is assumed that if two established nodes are already connected by an established road, but not along the geographical shortest path, a potential connection along the geographical shortest path will be eliminated from the candidate set if the total length of this connection is either less than 2 kilometers or less than 25 percent shorter than that of the current least-cost path between the two nodes. The rationale behind these rules is that expanding an existing road would be more efficient as compared to building a new road if the benefit of the new road in terms of travel time saving is marginal, as the former will incur less land acquisition cost. For example, the proposed connection $v_4 - v_5$ may be eliminated from consideration if the distance between v_4 and v_5 is only 2 kilometers or 25 percent shorter than that of the current connection $v_3 - v_5 - v_4$.

Second, estimate traffic demand on a proposed new road supposing it were built alone. The difficulty of this step lies in the trade-off between the accuracy of volume estimation and the running time required for evaluating a large set of candidate projects. For an internal connection, say $v_3 - v_4$ in Figure 10.2, the new road that connects v_3 and v_4 in opposite directions needs to be built with a length equal to the geographically shortest distance between v_3 and v_4, and a capacity of 400 vehicle/hr (in this study it is equivalent to one lane). A stochastic network loading is then performed once supposing the proposed road were already added to the network, and the loaded traffic on the proposed road is used to approximate the demand in equilibrium (estimation could be more accurate if network loading is repeated using MSA, but at a much higher cost of increased running time).

The flow estimation for an external connection follows the same procedure except that a network loading is not necessary, because the proposed new road on the external connection, if constructed, would serve as the only connection between G_m and G_n. Thus the flow on the potential road from G_m to G_n can be approximated as:

$$\hat{f} = \sum_{p_i \in G_m, p_j \in G_n} \hat{T}_{ij} \tag{14.28}$$

Third, estimate benefits from a candidate project. The increase in accessibility from a place is estimated as:

$$\Delta A = \sum_i W_i \sum_j J_j \left(e^{-\theta \widehat{l_{ij}}} - e^{-\theta \overline{l_{ij}}} \right) \tag{14.29}$$

For a project under centralized control, the gains in benefits from increased accessibility at all places will be summed up, while for a project under decentralized control, a local jurisdiction considers the increased value of local properties as well as the estimated toll revenue from non-residents as benefits.

Fourth, estimate the cost for a candidate project of new construction. To increase realism with the presence of lumpy investments, suppose a road is always designed in discrete lanes, and each lane represents an equal capacity of 400 veh/hr. The least number of lanes will be chosen such that:

$$C_a \leq \hat{f}_a/Q^* \tag{14.30}$$

The cost of constructing this amount of capacity will be estimated accordingly.

Fifth, select up to one project for new construction at a time. Under centralized control, only candidate projects with a benefit-cost ratio above one will be considered, among which the candidate road with the highest benefit-cost ratio will be built first with proposed capacities on both directions. The benefit-cost ratio of a candidate project is estimated as:

$$H_p^c = \pi \Delta A_p^c / E_{n,p} \tag{14.31}$$

Under decentralized control, matters become more complicated as a road project may involve multiple jurisdictions with autonomy. Given the complexity this process involves, heuristic assumptions are included in the implementation procedure. To approximate the cross-jurisdictional spillovers and spatial monopoly effect discussed in Chapter 13, we assume that a local jurisdiction is eligible to participate in a new construction project if this place is immediately connected to the proposed new road at its either end, or if the end node is not an established place but the jurisdiction is the (full or partial) owner of an immediate feeder link to this node. It is also assumed that the toll revenue and construction cost of a project will be split among participating jurisdictions in proportion to the projected traffic they will generate on the proposed links. To build the internal connection $v_3 - v_4$ in Figure 10.2, for example, eligible providers include the jurisdiction of v_3 and the owner jurisdictions of links $v_1 - v_4$ and $v_5 - v_4$ (as v_4 is not an established place). Additionally, each participating jurisdiction will estimate the proportion of resident travelers on the proposed road, and considers toll revenue only from non-resident travelers as its benefit. Suppose the planning horizon of a road project is N years, the estimated proportion of resident-travelers from place i on road p is λ_p^i, each jurisdiction then calculates the benefit-cost ratio of a project under its own interest:

$$H_i^p = (\pi \Delta A_i^p + (1 - \lambda_p^i)\chi_p^i R_p^d N)/\chi_p^i E_{n,p} \tag{14.32}$$

A participating jurisdiction will quit a project if the estimated benefit-cost ratio is below one. When a jurisdiction quits, the benefit-cost ratios of the remaining participating jurisdictions need to be re-calculated since the splits of toll revenue and construction cost change accordingly. Eventually, a list of candidate projects is developed with each participating jurisdiction in each project having an estimated benefit-cost ratio above one. Each jurisdiction then ranks candidate projects according to their benefit-cost ratios according to its own calculation. A jurisdiction would build the most cost-effective projects first, but it also compromises its own priorities to reach possible cooperation with other jurisdiction, as each participant has equal right to veto a project. In this case, negotiation between participating jurisdiction is in order. A project selection procedure implements this process as follows:

- Step 0: For any time period, starting from $n=1$.
- Step 1: For a list of candidate projects, each eligible jurisdiction considers only the top n candidate projects ranked according to their benefit-cost ratios.
- Step 2: If any participating jurisdiction of a candidate project ranks this project beyond its top n candidate list, this project will be eliminated.
- Step 3: If no qualifying project is found, $n := n + 1$ and go to Step 1; if n reaches a threshold (say 10), the whole process is terminated with no project selected; otherwise go to Step 4.
- Step 4: A qualifying project can be self-financed if the current balance of each participating jurisdiction can cover its share of construction cost. Among the qualifying projects that can be self-financed, the one with the lowest financial risk (indicated by the ratio of anticipated toll revenue to construction cost) will be built. If no qualifying project can be self-financed, a loan is involved. The bank agent will instead take over the decision power of project selection and minimize its own financial risk by selecting the project with the lowest ratio of projected toll revenue to the total amount of loans.

14.3 Simulation experiments

Now that the model is constructed, it is tested with a pre-specified set of established places on a hexagonal complete graph[3] in two experiments: Experiment 1 is executed under centralized control by which a central government builds and manages all the roads, while Experiment 2 implements decentralized control under which local jurisdictions build and operate roads on their own. It is assumed that one local jurisdiction represents one and only one established place. Hexagon centers represent the possible locations of places, from which ten locations are randomly selected. Two of them are established as bigger places each with 50,000 residents and 50,000 jobs, while the remaining eight ones are much smaller ones each with 5,000 residents and 5,000 jobs. The distance between two neighboring hexagon centers is specified as $\sqrt{3}/3$ kilometer. The initial set of places is illustrated in Figure 14.4(a), where dark dots represent established places while gray dots and lines represent potential nodes and links.

---

[3] The hexagonal graph is created following Haggett (1966), in which vertices and edges represent the possible location of established nodes (places or intersections) and established links (segments of roads), respectively.

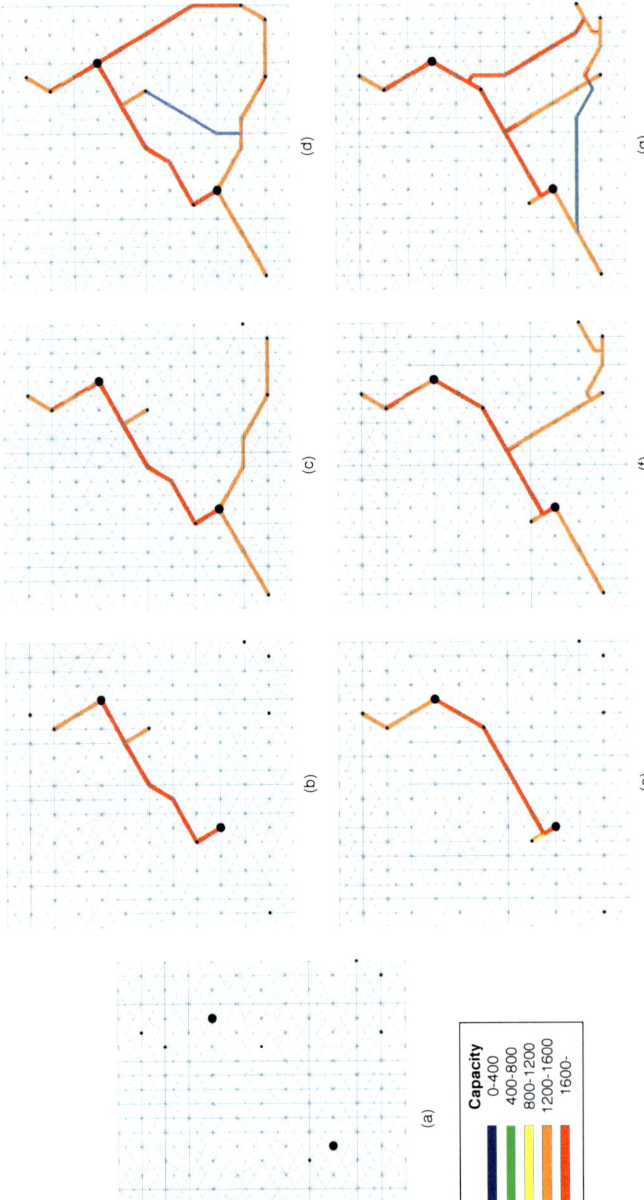

Fig. 14.4 Snapshots of incremental network growth (a) the initial network; (b)-(d): iterations 4, 8, 12 of Experiment 1; (e)-(g): iterations 4, 8, 12 of Experiment 2

Table 14.3 lists model coefficients and their specified values for the experiments. These models enable the examination of the sequential deployment of a road network in a controlled environment, and the evaluation of resulting system performances with different pricing and investment policies under alternative governance structures. While it has the potential to assist with policy making in practice, it needs to be noted that the present version of the models is not fully calibrated, and includes a number of arbitrarily specified parameters. In this regard measures of effectiveness (MOEs) presented below provide more of a numeric demonstration.

Table 14.3 Specified values for simulation experiments

Parameters	Value	Source
N	25 years	Common practice
υ	$10 /hr	Common practice
r_s, r_l	0.03, 0.05	Specified
κ	0.5 veh	Specified convergence rule
t_0	1.0 min	Specified
ξ_0, ξ_1, ξ_2	0, 0.25, 0	Specified
ψ_0, ψ_1, ψ_2	0, 0, 0.25	Specified
Q^*	0.95	Specified
ρ_1, ρ_2	0.2, 0.75	Behavioral assumptions on heuristic price probing
ω_1, ω_2	-30.6, 9.8	Empirical estimates by Zhang and Levinson (2005)
θ	0.048/min	Empirical estimate by Levinson et al. (2007)
$\beta_0, \beta_1, \beta_2$	5.79, 0.50, 0.39	Empirical estimates by Levinson and Karamalaputi (2003a)
ϕ	1/0.11	Adopted from Suwansirikul et al. (1987)
π	$12.71	Empirical estimates by El-Geneidy and Levinson (2006)

In Experiment 1, new construction ends after 11 years adding 11 two-way roads to the road network, and expansion continues until the simulation is terminated at the end of the 50th year. The first nine connections are external connections while the last two are internal connections. Eventually the travel demand on the established network reaches a total travel time of 50.1 billion vehicle hours or a total travel distance of 291 million vehicle kilometers annually. As shown in Figure 14.5, a test compares the predicted volumes (estimated in the investment models before the proposed roads are actually built) on the new roads versus their assigned volumes (estimated in the travel demand models next iteration after the roads are actually constructed) over the first eleven consecutive years (to be succinct, the volumes on only one direction of new roads are depicted). As can be seen, the new construction model performs well in predicting the travel demand on a proposed new road before the road is actually added to the network.

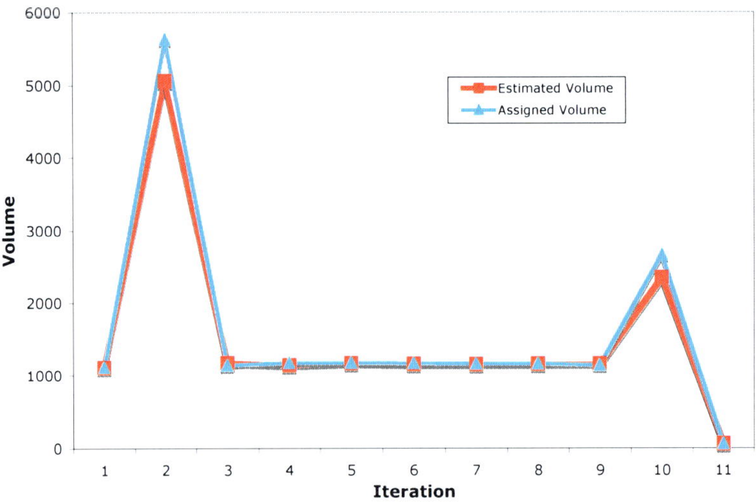

Fig. 14.5 Estimated versus assigned volumes on newly constructed links in Experiment 1

In Experiment 2 new construction also occurs during the first 11 consecutive years, while the expansion continues until the end of the 50th year, when an annual travel time of 54.0 billion vehicle hours or a total distance of 299 million vehicle kilometers is consumed.

Figure 14.6 compares the resulting road investment in Experiments A and B in terms of their (a) cumulative spending on new construction (b) cumulative spending on expansion, (c) cumulative toll revenues, and (d) spatial accessibility. As can be seen, local jurisdictions under decentralized control invest more in road infrastructure than the central government under centralized control, with a little bit higher spending on new construction and much higher expansion expenditure. The road networks developed in the centralized and decentralized scenarios, however, eventually provide rather close spatial accessibility, which implies centralized road provision is more efficient as opposed to decentralized provision when the central government has perfect information on individual links and the costs associated with bureaucracy, operation, and management are ignored.

Centralized pricing policies based on marginal cost pricing result in a deficit of 1.58 billion dollars at the end of the 50th year. Note that the empirical investment models specified in Equations 14.23 and 14.24 with β_1 smaller than 1.0 imply economy of scale in capacity. This is in agreement with the marginal cost pricing theory which predicts toll receipts will fall short of the facility costs if there are economies of scale in road provision (Gómez-Ibáñez et al., 1999). Additionally, even though jurisdictions might accurately predict the demand and revenue on a proposed new road for the near future when the investment decision is made, as more and more new roads are built in the long run, the demand on previously built roads may drop

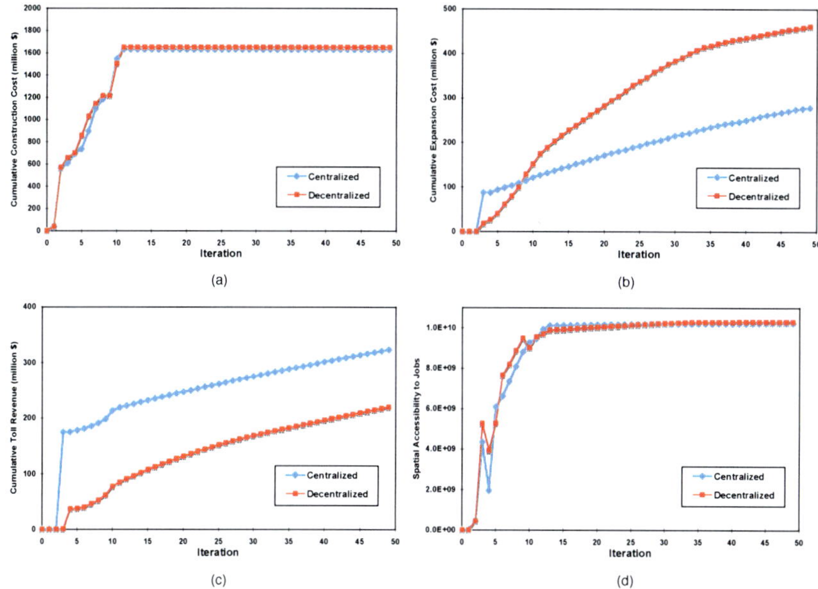

Fig. 14.6 Comparison of MOEs between Experiments 1 and 2

(with no demographic changes assumed) and the profitability of previous projects be undermined, thereby leading to financial deficit on road infrastructure.

Under decentralized control, local jurisdictions as a whole collect much less toll revenue than in the centralized scenario, despite the fact that they invest a lot more in road infrastructure and attract more traffic, which together results in a larger deficit (1.89 billion dollars) under decentralized control (though it would be expected that toll revenue will eventually cover all the investments as time goes on, since large-scale investment has been completed). Less toll revenue is collected under decentralized control probably because local jurisdictions have to compete with each other on parallel routes for through traffic. Another observation worth noting is that local jurisdictions build new roads faster than the central government. This can be explained by the fact that when local jurisdictions evaluate a road project, they consider not only the increase of spatial accessibility, but also the future toll revenue from non-resident travelers as the benefits.

Starting from the same network shown in Figure 14.6(a), the staged network growth under centralized and decentralized jurisdictional control is illustrated in Figures 14.6(b)-(d) and in Figures 14.6(e)-(g), respectively. As can be seen in the Figure, the sequences of road additions in both cases are similar: in the first four iterations, a central corridor is built between the two bigger places, forming the backbone of the road network; in the next four iterations, minor places in the corners are connected one by one; internal connections are then added in the last four iterations, which create rings and serve as shortcuts between places. The difference between the two scenarios lies in the fact that local jurisdictions, rather than directly connecting to other places, tend to connect to a node in the middle of a link so that the construction cost could be split among more involved places. Consequently, different network topologies emerge under centralized versus decentralized control. As can be seen, centralized control results in a network that connects places on a belt, on which roads are complementary to each other, while under decentralized control places on the south-east corner are connected to others via two parallel roads built by different groups of jurisdictions, which are competitors to each other.

14.4 Findings and concluding remarks

Road infrastructure evolves over time in a complex process as part of a dynamic and open system including travel demand, land use, as well as economic and political initiatives. This chapter represents the deployment of a road network as an endogenous process of incremental connections in a simulation environment, which is played out as the outcome of the decisions made by travelers with regard to destination and route choice in the short run, and by central or local governments with regard to pricing and investment policy decisions in the long run, under different political initiatives.

This study finds that demand-performance equilibrium can be reached on a centralized network between road suppliers, operators and travelers if a link-by-link marginal-cost pricing policy is adopted. In the decentralized case, a heuristic pricing-probing measure is proposed to estimate toll rates set by local jurisdictions on their subordinate links. This heuristic method deserves further investigation in terms of how independent road operators will actually adjust their toll rates in response to the variable demand with incomplete information on the competitor and cooperator links.

Simulation experiments disclose that both centralized and decentralized road provision resulted in a financial deficit. This finding agrees with the marginal-cost pricing theory in that toll receipts will fall short of the facility costs if there exist economies of scale in road provision. The inefficient road investment may also be attributed to the myopia of an incremental investment process that relies intricately on a sequence of network growth decisions, even if jurisdictions pursue an optimal investment strategy at the point when the decision is made. Simulation results also demonstrate that under decentralized control jurisdictions tends to make a larger investment while collecting less toll revenues on road infrastructure as compared to a central government, suggesting centralized road provision is more efficient in the presence of positive spillover effect, if the central authority has complete information on individual links and if the inefficiencies associated with its bureaucracy, management, and legislative decision-making process are ignored. In reality, the trade-off between the factors favoring centralization (such as economy of scale and economy of scope in road provision) versus those favoring localization (such as improved information on local markets) involves more complexities, and whether they could be fully addressed in simulation deserves further investigation.

The SONIC/GC model developed in this chapter has the potential to answer questions why urban road networks have developed into various topologies, and how decision makers could guide the development of transportation networks into their desired direction, especially when facing the choice of different pricing and investment strategies under alternative ownership structures.

Chapter 15
Forecasting

15.1 Introduction

A core problem of transportation planning is to identify infrastructure projects in which scarce resources are invested to maximize the provision of the public good. Some agencies proactively develop comprehensive transportation plans to guide these decisions and to provide certainty for other agents in the urban system, others make decisions by reacting to evolving market conditions and travel demands. Whether there is a comprehensive plan or not describing the "final" state of the network, the timing of future investment decisions is rarely specified beyond the current (typically six-year) Capital Investment Program.

From the late 1950s through the 1980s, the Minnesota Department of Transportation (Mn/DOT) and other state transportation agencies focused primarily on the construction of the US Interstate Highway System. Mn/DOT relied on the nationally developed Interstate Plan and the locally developed Backbone System Plan to guide this effort (Minnesota Department of Transportation, 2001). After completion of the Interstate, focus shifted within transportation agencies throughout the US from large-scale capital-intensive investments to the improved management of a mature infrastructure and an increased concern for the environment.

Policy plans in the 1970s and 1980s aimed to complete the metropolitan Interstate Highway System. Because the system was smaller and still new, the focus on management and preservation in those plans was not nearly as great as today. By the mid-1990s, the excess roadway capacity built in previous decades was largely utilized during peak periods, and problems with levels of congestion started to rise in the metropolitan area (Minnesota Department of Transportation, 2001). Non-recurring congestion has increased as well, and it was found that 13% of traffic crashes were secondary crashes from incident-related congestion (Minnesota Department of Transportation, 2005*b*). Without any excess roadway capacity, safety issues rising in prominence, and new budget constraints, the need for better planning strategies arose.

Decision-making for investing in transport infrastructure is complex and political as well as technical, thereby holding intrinsic interests for researchers from various fields. Political economists have shown a long-lasting interest in the provision of public roads under different levels of jurisdictional controls (Oates, 1972; Knight, 2002; Besley and Coate, 2003). Chapters 13 and 14 examined governance issues in transportation provision. In transportation planning studies, the prevalence of travel demand forecasting models (Sheffi, 1985; Ortuzar and Willumsen, 2001) since the 1950s made it possible to forecast travel demand on networks based on user equilibrium, thereby allowing traffic flows to be incorporated as an endogenous factor in forecasting network growth. Notably, network design problems (NDPs) develop a general bi-level framework in which the upper level represents the investment decision-making of transport planners to optimize social benefits within constraints based on the equilibrium flow pattern obtained from the lower level (Yang and Bell, 1998). Although NDP provides an effective tool to predict changes to networks (Davis and Sanderson, 2002), it fails to consider jurisdictional initiatives in the decision-making processes.

In order to gain a better understanding of the jurisdictional decision-making processes in urban planning practice, Montes de Oca and Levinson (2006) interviewed the planners, engineers and staff from Mn/DOT, the Metropolitan Council, seven counties comprising the Twin Cities metropolitan region, and the City of Minneapolis, the largest city in the region. Official or *stated* decision rules of different jurisdictions were outlined, disclosing that road projects are prioritized for federal or local funding mainly based on their safety records, pavement conditions, level of service, and capacity. On the other hand, Levinson and Karamalaputi (2003*a,b*) adopted a statistical approach to examine the expansion and new construction on a road network. They, after estimating statistical models on two decades of data from the Twin Cities, revealed that the likelihood for the expansion or new construction of a link is associated with a range of factors such as the present conditions of the network, traffic demand, project costs, and a budget constraint. Decision rules developed based on the statistical results are referred to as *revealed* rules in this study. As part of this research, the stated and revealed decision rules will be described later in detail.

This research investigates the timing and location of transportation investments in the seven-county Minneapolis-Saint Paul metropolitan area in Minnesota, and evaluates the effects of the investments in shaping the Twin Cities' road network. Stated and revealed decision rules of existing link expansion and new link construction developed in previous studies are included in a simulation suite to forecast the growth of the network based on the present and historical conditions, traffic demand, demographic characteristics, project costs and budget. The endogeneity of the network structure is a key contribution of this research over previous forecasting analyses that consider the topology of networks as exogenous.

In the next section, the overall model and research method is described. The following section overviews the simplified travel demand model that is constructed and calibrated using the historical data of the Twin Cities, which provides a research platform on which alternative scenarios may be tested. Then a set of investment models: budget, cost, and investment prioritization, are developed using the empirical data. The models are applied and alternative scenarios (forecasts) are produced and compared. The chapter concludes with what was learned from undertaking this modeling exercise as well as recommendations for future analyses.

15.2 Model

As part of this research, System Of Network Growth for the Twin Cities (SONG/TC) is developed on an open-source Java platform. A flowchart illustrates the model framework in Figure 15.1. While it looks similar to the abstract models presented in earlier chapters (SONG, SOUND, SONIC, and SIGNAL), this model is enhanced in many respects. In particular, this model, rather than assume independent investment rules on individual links, includes two alternative sets of "stated" and "revealed" investment rules, which are more representative of realistic investment decision making processes in metropolitan transportation planning. Moreover, its component models, such as travel demand forecasting models, budget models and cost models, are fully calibrated using the historical data in the Twin Cities.

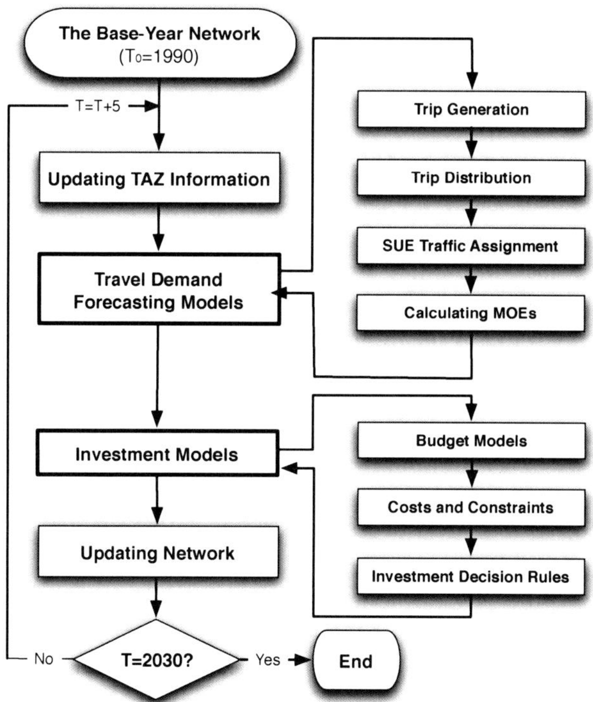

Fig. 15.1 Model framework of SONG/TC: System Of Network Growth for the Twin Cities (SONG/TC)

As can be seen, the program implements the growth of the Twin Cities road network as an iterative process at five-year increments. Each iteration includes four steps: demographic information updating, travel demand models, investment models, and network updating. Provided that it takes around 30 minutes to run one iteration of the program (dual 2GHz PowerMac G5, 1GB DDR), a five-year increment will substantially save the computational time as compared to smaller simulation periods (e.g. one year).

The first 15 years (1990-2005) can be thought of a warm-up phase for the calibration of the travel demand models. The network geometry, link attributes, and demographic information in the years of 1990, 1995, and 2000 are obtained from the Metropolitan Council's planning models as exogenous inputs to this program.[1] The investment models are skipped during the calibration phase (so that the travel demand models can be calibrated) and the network is updated with exogenous link capacities at the end of each five-year period. In calibration, model parameters were adjusted to minimize the difference between the predicted volumes of 2005 on major highways against the actual traffic counts obtained via loop detectors.

As the simulation continues with calibrated travel demand models, the forecasting phase starts from 2005 with the exogenous network topology, the estimated flow pattern (traffic across links) from the last simulation period (2000-2005), and the exogenous TAZ information provided by the Metropolitan Council. The Metropolitan Council also provides the forecasts of demographic information for the internal TAZs in the Twin Cities region every decade to 2030. The demographic forecasts every five years are then estimated by interpolation. For the 35 external stations, as traffic counts are available only for 1990 and 2000, it is assumed the volume of each external station will increase at a compound rate of 2% every year since 1990, consistent with the average rate calculated from 1990 and 2000 actual traffic counts at external stations. At the beginning of each five-year period from 2010 through 2030, the exogenous demographic forecasts are input to update the TAZ information, which is followed by the travel demand models, described in the next section, producing as outputs a network flow pattern and measures of effectiveness (MOE); these results are inputs to the investment process, which requires budget estimates (in part determined by vehicle distance traveled, as revenue depends on the gas tax) and cost estimates of potential projects, ranks potential discrete improvements (separately for the state and each county), and funds the highest ranked projects until the separate budgets are exhausted (once there is no budget available, there is a leftover deficit for the next time period); the projects upon implementation will change the network topology, which is updated endogenously before the time period is incremented.

---

[1] The original transportation planning network provided by the Metropolitan Council comprises 20,380 links, 7,723 nodes, 1,165 transportation analysis zones (TAZs) in the seven-county Metro Area, and 35 external stations. The planning network was modified to accommodate potential but unbuilt links, in particular what we call *legacy links* (projects that are in old transportation plans from the 1960s but that have not yet been constructed). The general idea is that if a legacy link intersects an existing link, there is a creation of a new node and the old link is divided into two different links. In the revised network representation there are 20,398 links and 7,733 nodes.

15.2.1 Travel demand models

The travel demand models include three component models of trip generation, doubly constrained trip distribution, and Stochastic User Equilibrium (SUE) traffic assignment, which simplify the traditional travel demand forecasting process by dropping mode choice, and instead directly estimate vehicle trips. We also do not model freight trips directly, and instead inflate passenger car trips to account for missing trucks. We are modeling traffic in the AM Peak Hour (the average hour between 6:00 am and 9:00 am), calibrating against that, and then using peak hour to daily expansion factors where required to obtain Annual Average Daily Traffic or AADT (which is required in some of the investment models). Peak hour volumes rather than AADT are used for calibration because peak hour volumes are accurately measured by detectors on a continuous basis and well maintained by Mn/DOT's Traffic Management Center, while many AADT measures are just estimates of actual traffic volumes. The calibration managed to reduce the average error between predicted and observed peak hour volumes on all major highways in the Twin Cities region to 0.78%, and the root mean square error (RMSE) to 30.0%.

The models essentially simplify the transport planning models developed by the Metropolitan Council. While the Metropolitan Council models capture more details about certain aspects of travel and can be more accurate, it is at the cost of requiring more data and resources. This research examines the investment on road networks and its effects by modeling multiple years of network growth, which is computationally more intensive. To do that in reasonable time, some details in travel demand forecasting are sacrificed.

It is worth noting that the computation for the initial year (1990) is iterated (using method of successive averages) between inputs to trip distribution and outputs from route assignment to obtain an equilibrium. Trip distribution requires peak hour interzonal travel costs (C_{ij}) as input, which are the output of traffic assignment. This is particularly important for the base year where we do not have a congested seed travel time matrix a priori. In this research, the initial network flow pattern is estimated by running the program beginning with free flow travel times on the 1990 network geometry, and iterating between trip distribution and route assignment (using outputs of assignment as inputs to trip distribution) until the maximum difference in travel time between two successive iterations is 0.1 hours for any OD pair, and the average difference is .0025 hours for all OD pairs. Once these criteria are met, the resultant flow pattern is used as exogenous input for the base year network and no distribution and assignment iteration is undertaken during the subsequent simulation periods.

As the focus of this research is the stated versus revealed investment process of network growth, the detailed description of the component models is skipped. More details on model setup and calibration can be found in Levinson et al. (2006). The parameters with specified values for the travel demand models are summarized in Table 15.1.

Table 15.1 Parameters and their specified values in the travel demand models

Parameter	Value	Description
α, β	0.15, 4.0	Coefficients in the BPR function (Bureau of Public Roads, 1964)
κ	0.11	Scaling Factor that converts daily traffic to peak hour traffic, adopted from Suwansirikul et al. (1987)
ϕ	20%	Calibrated percentage increase of initial highway capacity, based on the assumption that the real capacity on highways is underestimated by Metropolitan Council.
ε	0.048/min	Calibrated travel cost friction factor in gravity-based trip distribution.
θ	0.2/min	Scaling factor in stochastic route choice, adopted from Davis and Sanderson (2002). A value of 0.2 indicates that if one route is 5 minutes faster than the other, 3 out 4 travelers will choose the faster route.

15.2.2 Investment Models

The investment models predict budgets, estimate costs and constraints of potential road projects, and apply investment decision rules. This process outputs changes to the network in terms of link addition and link capacity expansion. The implementation of the investment models is illustrated in Figure 15.2 and the component models are discussed below.

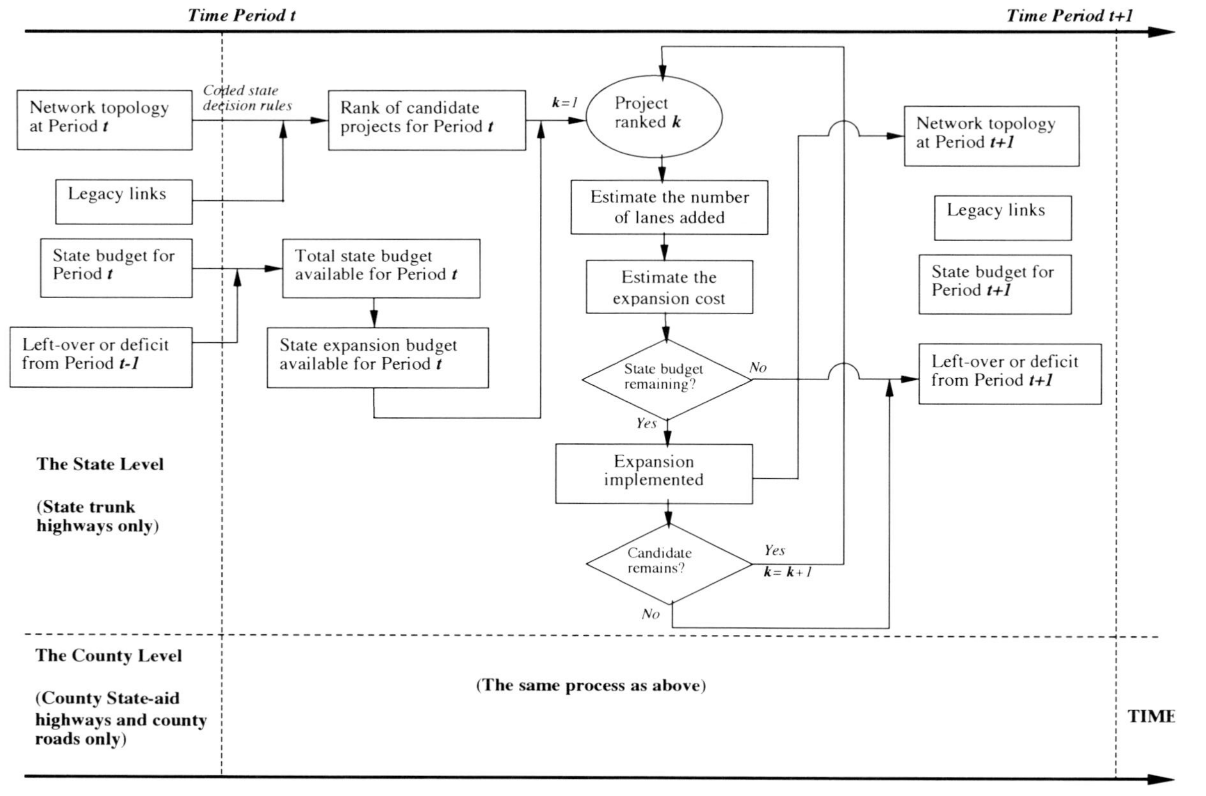

Fig. 15.2 Flowchart of the investment process

15.2.2.1 Budget models

In order to predict how much construction will occur in a given year, there is a need to know the available transportation budget.

Links that belong to the state (including Interstates, US Highways, and state highways) are ranked by state rules and are constrained by state budgets. Links that are under county jurisdiction are ranked by the respective county level decision rules. These links include County State Aid Highways (CSAH) and county roads. Other links that are not owned by these jurisdictions (such as park roads or roads owned by cities or townships) are not modeled in this investment model, and are assumed static. For this reason there is a need to estimate two different budget models for the state and for counties.

Once all links have been scored under each jurisdiction's decision rules, links are sorted and the budget is allocated to the links with the highest scores. A general assumption is that jurisdictions will spend all their budgets in that time period. If budgets are short on building one last project, jurisdictions will borrow from the next time period (decreasing available revenue in that subsequent period).

In order to predict the budget more accurately, the vehicle kilometers of travel (VKT) numbers predicted by this program are adjusted to published ones obtained through public agencies based on measured and estimated traffic counts. This is because the planning network used in this research does not include every link belonging to a particular jurisdiction, meaning that VKT produced by this model may underestimate the real counts.

The State budget model is estimated by regressing expenditures on state managed routes (Interstate, U.S. and Minnesota Trunk Highways) made by Mn/DOT. No distinction is made between the sources of funding. The regression model takes into consideration data available for the years from 2000 to 2004. A variety of regression models were tested, they included population, annual growth, residential density, network size, number of crashes, pavement conditions, households, income per household, car ownership, year, households per population as predicting variables. However, the simplest model, determined only by VKT (vehicle km traveled only on state managed routes), proved to have the greatest explanatory power. As displayed in Table 15.2, this model produced an r-squared of 0.82. Primarily through gas taxes, each vehicle km traveled adds approximately 0.6 cents to the state road budget. There were a total of 35 observations (seven counties by five years each).

The county budget model is estimated by regressing the expenditures made by the counties on County State Aid Highways and county roads. A number of variables were tested, including population, annual growth, residential density, network size, number of crashes, pavement conditions, households, income per household, car ownership, year, households per population, and shortest distance from the zone's centroid to either Minneapolis or St. Paul downtown. The model was estimated based on 28 observations for the years 1990, 1995, 2000 and 2003 (seven counties by four years each) and the final model that provided the highest r^2 (0.92) with significant variables is also presented in Table 15.2. Note that the predictive value of the model may be hampered by the yearly dummies; due to the limited number of

Table 15.2 Highway budget models

	State of Minnesota		Counties	
	Coef.	t Stat	Coef.	t Stat
Intercept	22146206	5.55	10289970	4.04
Households			-78	-1.98
VKT	0.006	12.39	0.0047	4.16
Dummy for 1990			-11552700	-3.64
Dummy for 1995			-5900211	-2.03
Dummy for 2000			-3491842	-1.32
Adjusted R Square	0.82		0.9	
N	35		28	

records available, however, we did not test the stability of the model structure over time. Instead it is assumed the estimated spending pattern is stable in the forecast years.

Transportation budgets need to be separated into maintenance and construction budgets. Based on the current spending pattern in the Twin Cities Metropolitan Area as noted in Minnesota Department of Transportation (2001), this research assumes 21% of the total budget is spent on construction and 79% is spent on maintenance. Sensitivity analyses are also undertaken (later in Scenario 5) to test the effects of varying the available construction budget. The construction budget further allocates funds between capacity expansion of existing facilities (which largely serve existing needs) and the construction of new facilities (which open up new areas to development).

A major modeling issue is the allocation of the 21% of budget devoted to network construction between expanding existing links and building new links. The number of existing links is known, as are their attributes (congestion level, crash rate, etc.); possible future links (new construction) on the other hand comprise a much more challenging problem.

Only a few *legacy links* in this analysis have already been clearly laid out. These legacy links have appeared on state maps and plans since at least the 1960s, and have been political promises to the affected areas that a new road would eventually be built. In the Twin Cities, state-level legacy links include the extensions of freeways Highway 610 and Highway 212. Figure 15.3 shows all of the links that were proposed in the 1960 Metropolitan Transportation Study that were (a) proposed and built, (b) proposed and not built (marked in red), and (c) not proposed at the time, but built (links that were not proposed and not built cannot be easily mapped).

For lower levels of government in the Twin Cities region, such long-term plans are uncommon. Yet from time to time, new links are constructed. Levinson and Karamalaputi (2003*b*) developed a series of rules that were used to identify potential links depending on the traffic at the nodes (which were assumed to already exist), length (not too short, not too long), and local characteristics (not crossing more

Legacy Links Map in Actual Network

Fig. 15.3 The map of legacy links in the Twin Cities

important links). That set of rules produced over ten thousand possible new links for the Twin Cities network, of which a few dozen were built in the past 20 years.

Since the rules for prioritizing expansions of existing links and construction of new links are different (which will be explained in Section 15.2.2.3), it is very difficult to compare them on a standard metric. One can compare two expansion projects or two construction projects, but there is no easy translation between them. Thus it is easier to establish separate budgets for link expansion and new link construction, rather than make them compete directly for resources. From 1978 to 2004, in the

Twin Cities Metropolitan Area, there were 945 lane km added to the transportation system at the level of principal arterial or higher. Among those, 821 lane km were new construction and the remainder were expansions of existing facilities. During those 26 years 85% of the dollars spent went to new link construction while 15% was allocated to link expansion and reconstruction. The average cost for new construction was $1,495,000 per lane km while for expansion it was $752,000 per lane km, which represents a 50% difference. As the network matures, this ratio is expected to change. For modeling purposes, we assume forecast expenditures will be split 50/50 between expansion and new construction (another split 75/25 is tested in an alternative scenario for comparison), until the number of new construction projects is exhausted, at which time, the excess new construction revenue will be reallocated to expansion.

15.2.2.2 Costs and constraints

This research adopts a model of facility construction cost estimated by Levinson and Karamalaputi (2003*a*). This model takes into consideration facility size, new construction (vs. expansion), road type, as well as the distance from the nearest downtown. This model was estimated on facilities that were actually built. It is important to mention that the cost model will underestimate costs for roads that were not built, for which high cost may have been a discouraging factor. One way to account for this is to better consider constraints on investment as additional costs. Alternatively, constraints can reduce the points allocated to potential projects in the decision rules (the approach we take). Two major constraints are available right-of-way (ROW) and environmental factors.

Interstates, highways, county roads and streets often require taking real property for ROW. This aspect needs to be addressed when analyzing results of the expansion/new construction of the possible transportation network additions. While in some areas there is a possibility of obtaining land on the side of existing roads to expand them if needed at a reasonable price, in many urban areas this is infeasible because of existing structures. The available ROW in the heart of urban communities is a constraint.

This research tried to consider the ROW available on both sides of the roads that are prospects for expansion using Geographic Information System (GIS) technologies. But there was no data available for this specific type of analysis. A GIS land use file including a category named "right of way" was available, but for this analysis more specific data was required (i.e. spatial location of each building within the parcel data, as well as specific location of highways within the ROW, lane width, sidewalk width, and so on).

There are significant terrestrial and wetland ecological areas in the seven-county metropolitan area to take into consideration for the predicted expansion of the transportation network. The areas are classified by Minnesota Department of Natural Resources (2003) as Outstanding, High, Moderate and Non-classified based on the importance of ecological attributes like size, shape, cover type diversity and adjacent land use. These areas include individual forests, grasslands and wetlands. Potential links that traverse these ecologically sensitive areas, as well as bodies of water like rivers and/or lakes and over parks as well, are marked as constrained.

The investment models with stated decision rules (discussed below) rank road projects by their scores that indicate their potential benefits, not costs. Costs are used to allocate available funds. Thus, when a link predicted for expansion and/construction is constrained by any of these areas, instead of allocating points, points will be taken away. Based on a scale 0 to 100, it is assumed constraints will cause 90 points to be deducted from that link, which in an era of constrained budgets, should ensure it does not get funded.

15.2.2.3 Decision Rules

Two classes of decision rules are used in the analysis: stated rules, garnered from interviews (Montes de Oca and Levinson, 2006), and revealed rules, determined by statistical analysis (Levinson and Karamalaputi, 2003*a,b*) and briefly described below.

For the stated rules, in some cases they are relatively clearly outlined in public documents, while in other cases, those processes were much more informal, and required judgment to ascertain. In order to uncover formal and informal procedures, performance measures, and decision rules that have been actually used, interviews were undertaken with Mn/DOT, the Metropolitan Council, County, and City of Minneapolis planners, engineers and staff involved in the decision-making process on future network growth. These interviews were conducted in groups as well as individually, in which the following free-form questions were asked:

1. What is the procedure for a project to be approved for construction?
2. What are the most important policies to look at when making decisions about a project for the network growth?
3. What are the main criteria to choose between different projects?
4. What performance measures are considered important when selecting a project?
5. Have there been changes in the criteria used today as those that were used 20 years ago about network development?
6. Are there any informal procedures for the decision-making process?
7. How important of a role do politics play on the decision-making process?

From these interviews a series of flowcharts were developed and encoded into the computer model. The decision flowcharts are made operational as If-Then rules. The If-Then rules implement a point allocation that covers the decision rules that are considered by each government jurisdiction in a numerical ranking format.

These points are assigned based on the characteristics of the roadways located in each county. Every county has its own decision rules. Counties that did not provide decision rules were assigned decision rules of a similar adjoining county. For decision rules that are based on subjective criteria (e.g. public support for a specific project), there was no numerical way to allocate points, therefore these types of decision rules were not taken into consideration for any of the calculations. The four main factors that were common between flowcharts were Safety, Pavement Conditions, Level of Service, and Capacity. Points need to be continuous in order to ensure that each project obtains a unique score from a jurisdiction, so discrete points associated with ranges were converted to continuum.

Taking Hennepin County as an example, the Normalized Scoring System of the county evaluates the funding needs of a road project from three aspects:

- Safety. Roads belonging to this county are categorized into groups according to road types (urban or rural, divided or undivided) and number of lanes. The ratio of the crash rate on a road to the average crash rate of its group is calculated based on which specific points are added to this road.
- Pavement quality. Pavement Quality Index (PQI) of each road is calculated and points are allocated to a road according to its normalized PQI
- Level Of Service (LOS). The ratio of Annual Average Daily Traffic (AADT) to current capacity is calculated as a LOS indicator for each road and points are allocated according to normalized LOSs. The decision rules developed by Scott County, on the other hand, are less structured, which consider only two criteria: if road location is among the top 200 high crash list, certain points are allocated to the road; if the ADT of a road is above a threshold, certain points are allocated; otherwise no points will be allocated. Note that crashes on roads are not endogenized in this program and historical crash records rather than predicted ones are used in applying the related decision rules in the forecasting process. Details on the flow charts, If-Then rules, and point allocation criteria are available in Levinson et al. (2006).

As the stated rules are primarily concerned with expansion projects, the revealed decision rules developed by Levinson and Karamalaputi (2003*a*,*b*) are also introduced, which apply the statistical models estimated respectively for expansion and for new construction. The expansion of facilities on the existing network by one or two lanes is estimated using a discrete choice model with independent variables describing conditions of the network, traffic demand, other demographic characteristics, estimated project costs, and a budget constraint. The likelihood of expansion of a link depends also on its upstream and downstream neighbors, as well as on the state of parallel links. The model suggests that high capacity links are more likely to be expanded. New highway construction was estimated in a discrete choice model to be based on the status of the network, project costs, the conditions on upstream and downstream and parallel links, and budget constraints. Algorithm was developed to generate a large choice set of potential new links, to which the discrete choice model was applied. New links providing greater potential access are more likely to be constructed.

15.3 Scenarios

A range of seven scenarios has been constructed to examine how the timing and location of road expansion and new construction predicted by the simulation program would be affected by varying (1) Decision rules for expansion, (2) Decision rules for new construction, (3) Total budget, (4) Budget split between expansion and new construction and (5) Choice set of potential new links. The scenarios are summarized in Table 15.3.

Table 15.3 Scenario descriptions

| | Decision Rules | | Budget | | |
No.	Expansion	New Construction	Total	Split of Expansion	New Link Choice Set
1	Stated	Revealed	Standard	50/50	Legacy
2	Most structured	Revealed	Standard	50/50	Legacy
3	Least structured	Revealed	Standard	50/50	Legacy
4	Stated	Revealed	4a 100%+ 4b 200%+ 4c 400%+ 4d 10%- 4e 25%-	50/50	Legacy
5	Stated	Revealed	Standard	25/75	Legacy
6	Revealed	Revealed	Standard	50/50	Legacy
7	Revealed	Revealed	Standard	50/50	All potential

Scenario 1 presents the baseline scenario, consistent with the assumptions described in last section. In Scenario 2, the most structured decision rules (those from Hennepin County) are applied for link expansion to every county; while in Scenario 3, the least structured decision rules (those of Scott County) are applied for link expansion. For new construction, the revealed decision rules are used to prioritize links in all cases.

In all the scenarios, the total budget available for road investment is estimated using the baseline budget model described in the last section, except in Scenario 4, where the budget alternatives are tested. Scenarios 4a-4e respectively assume the construction budget allocated to each jurisdiction for each time interval is increased by 100%, 200% and 400%, and reduced by 10% and 25% (which is essentially equivalent to varying the budget split between construction and preservation). The construction budget is split evenly between expansion of existing links and new construction, except in Scenario 5, where three-fourths of all dollars are allocated to new construction. When opportunities for new links are exhausted (all of the legacy links have been built), that budget is reallocated to link expansion.

In contrast to Scenarios 1-5 that adopt stated decision rules for expansion, Scenarios 6 and 7 adopt revealed rules instead. While all of the scenarios take existing links as a baseline and consider them for expansion, the scenarios differ in what links to consider for new construction. Consistent with the first five scenarios, Scenario 6 uses only legacy links as links that are eligible for new construction, while Scenario 7 adds a set of potential links that have not been pre-specified on historical maps due to the relative scarcity of legacy links that are available for investment.

The additional set of potential new links in Scenario 7 is generated using the model from Levinson and Karamalaputi (2003b). It begins by identifying all existing node pairs that meet a specific set of criteria. The type of potential link is identified based on the highest-level link coming into each of the nodes. If a node is attached to a freeway link, a potential new link will be part of the freeway link level. The potential links are constrained: new streets cannot cross existing higher-level roads (such as highways or freeways), but freeways and highways can cross streets. Every combination of two existing nodes is considered and the possibility of establishing a link between them is evaluated. The candidate link should be longer than 200 meters and shorter than 3200 meters in the Twin Cities area. Consequently, a total of 14,826 potential links are identified in the Twin Cities Metropolitan Area, though only a few of them are constructed each year according to the traffic condition and budget constraints.

Potential links in the additional set of candidates that would cross parks, water areas and other ecologically sensitive areas are excluded from the choice set (and were not constructed in the model). However, legacy links with such constraints are constructed with a penalty in length since the link has to detour in order to get built. This penalty length was assumed to be 1.4 times the airline length and consequently makes it more expensive to construct.

For all scenarios, once a link has been expanded, it is no longer taken into consideration for further expansion. For new construction, state roads are assumed to be two lanes in each direction, whereas county roads are only one lane in each direction. Newly constructed roads are eligible for expansion if necessary in the future. Additionally, it is assumed that all expansion and construction decisions for links are symmetric, which is typical in the Twin Cities Metropolitan Area. This means that in case of expansion and new construction an equal number of lanes will be added in both the ij and ji directions. However for one-way streets, only one-way expansion is considered, which allows asymmetric developments.

15.4 Results

Simulation results are summarized in Figures 15.4-15.10. To save space, only Scenarios 1, 6, and 7 are elaborated. For a full description of the seven scenarios and their respective results, refer to Levinson et al. (2006). Note in the figures that each simulation period is labeled by the first year in this period (thus 2005 represents the period from 2005-2010, 2010 represents the period from 2010-2015, and so on).

Figures 15.4 and 15.5 compare the scenarios based on overall results with respective regard to new construction and expansion for the state. New construction

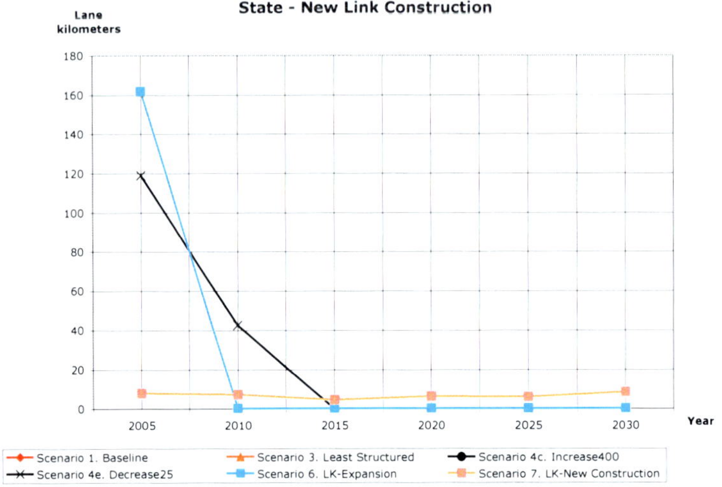

Fig. 15.4 New construction by the State for different scenarios

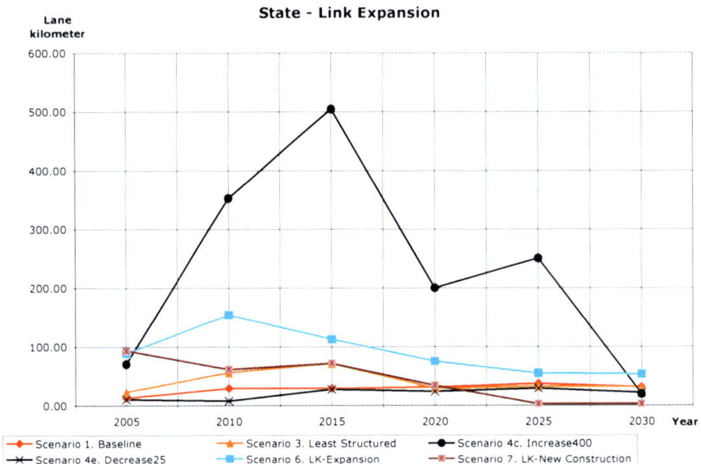

Fig. 15.5 Link expansion by the State for different scenarios

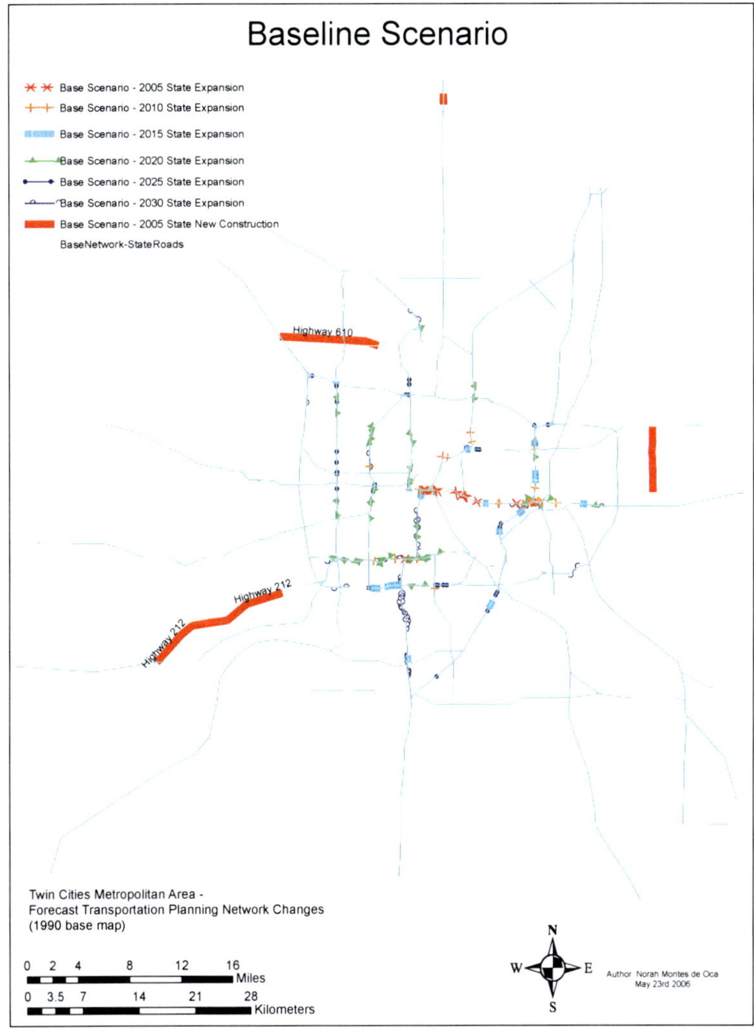

Fig. 15.6 Scenario 1. Stated rules Expansion + New Construction + Legacy Link Choice Set - State

plummets for all except Scenario 4e and Scenario 7 because all of the legacy links are built by 2010 (and the choice set of new construction is exhausted). Scenario 4e does not exhaust the new construction opportunities until 2015 because of a decreased budget, while Scenario 7 provides a larger set of potential unbuilt links to choose from. Once new construction opportunities are exhausted, the scenarios reallocate new construction funds to roadway expansion. Expansions are greatest in Scenario 4c which has a 400% increase in total budget available for construction. The other scenarios are much more similar in total amounts of expansion, though

Fig. 15.7 Scenario 6. Revealed rules Expansion + New Construction + Legacy Link Choice Set - State

Scenario 6 has more expansions than the others, because it adopts revealed decision rules which take into consideration economy of scale in expansion cost, so that longer roads can get expanded with higher priorities with lower cost rate per lane kilometer (Scenario 7, despite also adopting revealed rules, has fewer expansions because the money for new construction remains with new construction).

Figure 15.6 shows the predicted expansions in Scenario 1 (the baseline scenario) for the state. The model projects the state will construct continuations of Highways 610 and 212 in the 2005 period. In the 2015 period there will be some expansion on

Fig. 15.8 Scenario 7. Revealed rules Expansion + New Construction + Expanded Choice Set Model State

sections of I-35E and on I-494 west of I-35W. Sections of Highway 100, I-94, TH62 and I-494 show some expansion by the 2020 period. There will be some expansions as well on I-35W from south of I-94 to south of Bloomington by the 2030 period. Highway 10 will also have some expansions over time. There will be some other small expansions spread across the region as well. In the 2015 and 2020 periods the demand for new construction is in the northwest part of the metropolitan area.

Under Scenario 6 (Revealed Decision Rules, Restricted New Construction Choice Set), as shown in Figure 15.7, there is new construction at the same time as the base

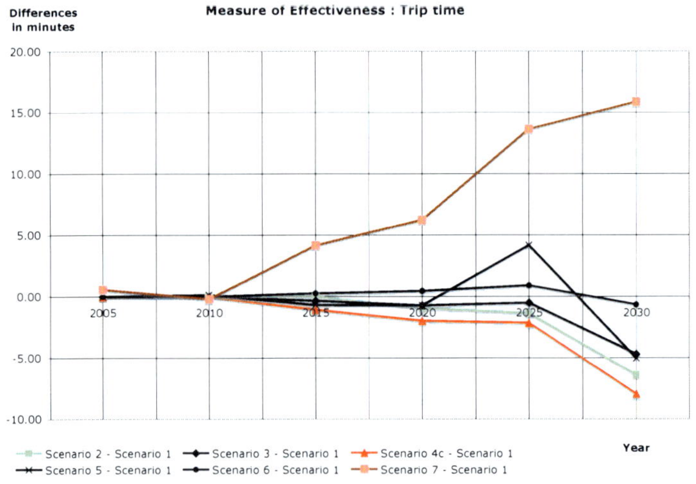

Fig. 15.9 Measure of Effectiveness - Trip time

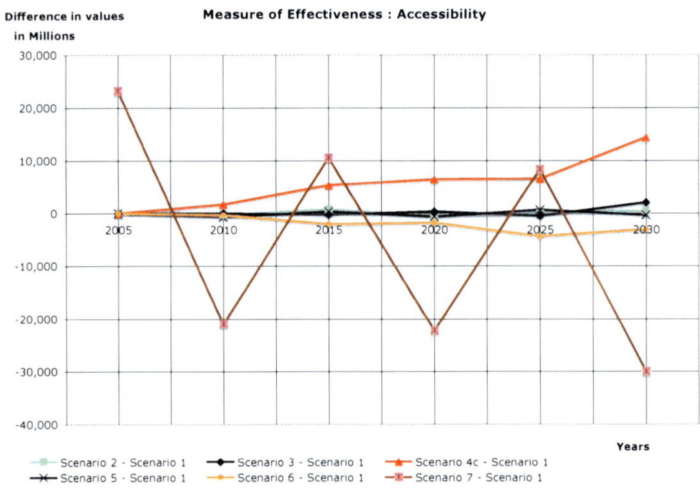

Fig. 15.10 Measure of Effectiveness - Accessibility

scenario suggested. For Scenario 7 (Revealed Decision Rules, Unrestricted New Construction Choice Set), shown in Figure 15.8, at the state level there is no clear pattern either for expansion or new construction. It occurs across the region in different years. While Scenario 7 would be investing in fewer than 10 lane kilometers of new projects per period, the investment would be continuous over 25 years, and not end after the set of legacy links was exhausted. By and large, these investments would not occur on the legacy links.

The program calculates a range of MOEs for each simulation period including average trip time, average trip length, vehicle kilometers of travel (VKT), vehicle hours of travel (VHT), and gravity-based accessibility. Levinson et al. (2006) includes the detailed calculation and predicted fluctuations for each of the MOEs. To illustrate the temporal change of MOEs due to road investment, Figures 15.9 and 15.10 compare the fluctuations of average trip time and accessibility, respectively, as predicted in the seven scenarios.

In the base scenario, the predicted individual trip time has increased from 15.0 minutes in 2005 to 37.8 minutes in 2030. Figure 15.9 compares individual trip time over years between Scenarios 2-7 against the base scenario. As can be seen, Scenario 4c (with a 400% budget increase) provides the shortest trip time among all scenarios in 2030. In contrast to all of the other scenarios that eventually improve upon the base scenario, Scenario 7 results in a trip time that is about 15 minutes longer than Scenario 1. While the relative positions of trip times across the scenarios are plausible, all the scenarios predict a steep increase in the trip time over years. This may be exaggerated (if historical changes are any guide) due to a variety of modeling assumptions, most significantly, the land use assumptions, which are not likely reasonable in forecast years, the lack of peak spreading in the model, the relative insensitivity in the gravity-based trip distribution model to changes in travel time, as well as changes in travel demand at external stations, for which we have a very simplistic forecasting procedure due to limited data.

The gravity-based measure of accessibility in this research evaluates the ease of reaching employment opportunities (retails and non-retails) at destinations, impeded by the generalized travel time from origins to destinations. The predicted accessibility in the base scenario increases from 1.015×10^{12} units in 2005 to 1.500×10^{12} in 2030. The increasing accessibility throughout network growth indicates the benefits travelers in the Twin Cities region have gained due to road investment. Figure 15.10 compares the measure of accessibility from Scenarios 2-7 to that in Scenario 1. Scenario 4c again achieve the highest accessibility in 2030 with its increased budget, indicating greater investment can achieve lower travel times and as a result higher accessibility. As compared to changes in total investment, switching specific decision rules or choice sets applied to the network make only marginal changes in network performance. While accessibility has improved substantially in all scenarios, Scenario 6 and Scenario 7 based on revealed decision rules for link expansion result in lower accessibility as compared to the other scenarios, suggesting stated decision rules perform better than revealed rules in terms of improving regional accessibility.

15.5 Findings and concluding remarks

This research, perhaps for the first time, forecasts changes in transportation networks as a function of empirically derived models, using a travel demand model based on economic theory and observed information. While one must treat with caution any specific results, the exercise is valuable.

This research finds that stated decision rules (in Scenarios 1-5) perform generally better than revealed rules (as adopted by Scenarios 6-7) in terms of improving regional accessibility by investing resources in road infrastructure. The reason may be two-fold. First, stated decisions vary by jurisdictions and may better reflect local investment needs while revealed rules are developed across the whole region and may be less flexible to the heterogeneity of jurisdictions; second, stated rules evaluate investment needs mainly based on present road characteristics, while estimating a statistical model for revealed rules requires a large data set across past years, for which some predictors such as level of service and crash rate are difficult to incorporate because of limited availability to historical data.

Another interesting observation is that decision-making with a limited choice set of legacy links for new construction (Scenario 6) appears to achieves a shorter trip time and higher accessibility as compared to that with a larger set of potential new links (Scenario 7). This may be for several reasons. One obvious reason is that more construction (both expansion and new construction) results in Scenario 6. Furthermore, supposing legacy links are strategic links identified in previous transport plans that can effectively complement the current highway network in the Twin Cities, the revealed decision rules didn't identify them among an expanded set of potential links and result in a less contiguous pattern of construction in Scenario 7 than in Scenario 6.

Extensions of this research from a modeling perspective can proceed in several directions. First, the step length between iterations can be reduced from a five-year model to a one-year model. One of the reasons for wanting to change from five-year model to one-year model is to test an evolutionary model of network growth. In the one-year model, only a fraction (say 20%) of all work trips change destinations in a given year. This means that 80% of trips in previous year would not change, only 20% of OD demand in that year and additional OD demand this year will be redistributed according to the congested travel time calculated at the end of the previous year.

One major criterion we were unable to model was pavement condition, due to a lack of geographically accurate and complete data on the current pavement condition across the regional network. Should this data become available, it would be useful to re-introduce this variable as a factor affecting the timing of investments. Another criterion to be included is safety. A crash rate model needs to be estimated and incorporated in the program to make crashes on roads endogenous. This represents a challenging part of future work as safety plays an important role in decision-making processes of jurisdictions.

Clearly, improvements can be made to the investment models, particularly in the way resources are allocated between new construction and expansion of existing facilities. The available information in those cases is different, resulting in different criteria used to prioritize those types of decisions. Additionally, better models of total revenue, and revenue available for investment, should be aimed for. Assuming a fixed share of total revenue is invested is unlikely, as the network grows and matures, we expect an increasingly large share would be associated with maintenance and preservation, though the data from the past decade do not point to any clear trends.

One of the great benefits of a modeling exercise such as that conducted in this research is not simply the predictions, it is the process, which requires coding decisions into a computer program in a logical way, forces the specifications of all of the assumptions that are often expressed vaguely in typical spoken and written human communication. There are many parts of the decision-making process that are underspecified in written documents, leaving ambiguity and opportunities for special-case politics rather than systematic consideration and evaluation of decisions according to agreed upon principles. While that ambiguity may be intentional, it reduces transparency in the system and opens it up for manipulation.

Part VI
CONCLUSIONS

This part highlights the findings from the preceding chapters, discusses their implications for evolutionary transportation planning, and indicates directions of future research.

Chapter 16
Retrospect

This book models and analyzes the growth of transportation infrastructure networks from an evolutionary perspective. While this evolutionary process is essentially driven by supply-demand interaction, this book undertakes the investigation with a main focus on the supply side of the story. The organization of this book is characterized by a series of stand-alone studies each examining an aspect of network growth. The integrating theme lies in their common interest in the bottom-up, decentralized investment decision-making processes that have driven network growth under interdependent interests.

Part II (Chapters 4-7) presented an *ex post* examination of transportation development using archived historical data. Chapters 4 and 5 examined the expansion of transportation networks as a discrete sequence of link additions using the historical data extracted from the skyway system of downtown Minneapolis and the interurban network of Indiana, respectively. Both studies provided empirical evidence on the proposition that the deployment of a transportation network has to some extent followed a predictable path by which accessibility is maximized. Depending on its ownership organization, though, supply decisions may be made under different interests to serve different groups of people.

Chapter 6 carried out an empirical analysis of network growth from a different perspective. In contrast with Chapters 4 and 5 that treated land use as exogenous, this chapter analyzed the causation effects in the coupled development of population and streetcars in the Twin Cities of Minneapolis and Saint Paul. It revealed that the rapid expansion of the Twin Cities electric streetcar system, rather than respond to the *existing demand*, has in large part been driven by other forces such as its technological superiority, monopolistic ownership, real estate development (which can be thought of as *prospective demand*), and people's reliance on the streetcar for mobility. Proximity to streetcar lines was such a crucial factor in the dispersion of residence that it had essentially shaped the suburbanization of the Twin Cities.

Chapter 7 investigated the existence or absence of first mover advantages (FMA) in the transportation sector. Four empirical examples illustrated qualitatively and quantitatively different aspects of first mover advantages in the development of modern transportation systems. The empirical case of London rails suggested the existence of first mover advantages in surface transportation networks, and revealed its close relationship with the spatio-temporal location of stations and network connectivity. Analysis of the global aviation system indicated a significant persistence of airlines at hubs, providing another instance of first mover advantages that are strengthened over time. The case of Twin Cities roads, on the other hand, implied that an accurate treatment of first mover advantages needs to consider both transportation and land use in an evolutionary context, for which the empirical approach is largely limited due to the scarcity of extensive historical data.

Part III (Chapters 8-10) explored the spontaneous organization of system properties that transportation networks globally possess, ranging from hierarchy and topology to sequence. Each of the chapters introduced an abstract, *ex ante* model which, tailored for different research questions, represents the evolution of transportation networks from innovative perspectives of self organization, degeneration, and incremental connections. Chapters 8, largely replicating the work of Yerra and Levinson (2005) and Levinson and Yerra (2006), constructed System of Network Growth (SONG) to evolve a transportation network based on completely localized investment processes, and demonstrated the spontaneous organization of link hierarchies in the network under a broad range of assumptions of initial conditions. Chapter 9 introduced the System Of Ultra-connected Network Degeneration (SOUND), which enabled a variable network in an iterative process of interaction, investment, and disinvestment. The model was validated on a historical data extracted from Indiana interurbans and executed on idealized network structures. Quantifiable indicators introduced for this analysis suggested the emergence of typical connection patterns such as hierarchies, rings, webs, hub-and-spokes, and cul-de-sacs, providing strong evidence for the spontaneous organization of network topologies. Chapter 10 presented System Of Network Incremental Connection (SONIC) to represent the chronological growth of a transportation network through a discrete sequence of link additions to the network between a pre-existing set of places, and demonstrated in a simulation that the model based on a heuristic "strongest-link" assumption associated with accessibility was able to reproduce the observed course of skyway additions in downtown Minneapolis to a statistically significant extent.

Extending the efforts in Part III, Part IV (Chapters 11-12) presented two additional models of network growth while endogenizing the evolution of urban space simultaneously. Chapter 11 modeled the diffusion of a transportation network coupled with place formation in System Of Network Incremental Connection and Place Formation (SONIC/PF). In complement to the empirical analyses in Chapter 7, this model was employed to analyze first mover advantages and their contributing factors in a controlled environment. Simulation experiments supported the speculations that FMA in a transportation network arises from the locational advantages acquired by early deployed facilities, and exhibits a reinforced strength as the network diffuses over time and space. Simulation results also substantiated the effects of initial land use distribution and network redundancy on the extent of first mover advantages.

Chapter 12 explicitly modeled the coupled development of transportation and land use as a spontaneous process driven by simple, myopic, decentralized decisions made by travelers, business owners, and resident workers. A series of component models is encapsulated in System of Integrated Growth of Networks And Land-use (SIGNAL) in as simple a way as possible to capture the salient properties of co-evolution. Simulation experiments demonstrated the spontaneous organization of land uses and network capacities into a stable order regardless of initial conditions, and revealed that the agglomeration of land use is reinforced by network dynamics, while the concentration of network capacities is reinforced by the differentiation of land use gradients.

From a normative stand point, Part V (Chapters 13-15) constructed and applied network growth models in governance and planning studies associated with transportation policy-making. Chapters 13 and 14 studied the institutional structures that underly the provision of transportation networks. Focusing on governmental provision, Chapter 13 presented a game-theoretic analysis of centralized versus decentralized governance choice on a serial road network. It reveals that, depending on the trade-off between the benefits of internalizing spatial spillovers across local jurisdictions in a centralized spending system against the costs associated with opaque legislative decision-making processes, governance choice reflects constituents' collective spending preferences for infrastructure, and may spontaneously shift as the network improves over time. Chapter 14 constructed System Of Network Incremental Connection for Governance Choice (SONIC/GC) to represent the incremental deployment of a road network driven by different political initiatives. Alternative sets of pricing/investment policies under centralized and decentralized governance were tested in a simulation environment, and the resulting network topologies as well as system performances evaluated. This research provides an effective tool for policy makers to evaluate the long-term effects of different pricing and investment strategies on the development of transportation networks. Chapter 15 developed System Of Network Growth for the Twin Cities (SONG/TC) and calibrated it using the empirical data from the Twin Cities metropolitan area. The model was applied to predicting the timing and locations of transportation investments on the Twin Cities' road network between 2005 and 2030. This research specified the processes necessary to run the network forecasting models with various decision rules in a planning practice. It is found that given the relatively small amount of funds available for network growth in a mature system, alternative decision rules make only small differences in overall system performance, though they direct investments to very different locations and affect the trip time and spatial accessibility in a significantly different way.

As can be seen, this book has treated different aspects of network evolution using different approaches empirically, analytically, and in simulation. Empirical models only investigated the *ex post* development of transportation networks. Limited by the availability of extensive historical data, the empirical approach usually provides us a relatively "coarse" picture of transportation development. Empirical studies, however, analyzed network growth against a concrete backdrop of its people, geography, economy, and culture, thereby revealing an *ad hoc* story for each transportation system in examination. Theoretical models in general are mathematically more rigorous, and have their strength in precisely describing the decision-making processes of network growth and their inter-connections. In order for mathematical tractability, however, theoretical exploration is usually limited on small networks in fairly simplified contexts. Simulation models excel in representing and analyzing the evolutionary process of network growth in a controlled environment. Agent-based simulation, in particular, enables the evolution of complex networks in a holistic, bottom-up process based on decentralized yet interdependent decisions made by stakeholders. A sound simulation, however, requires rigorous calibration and validation which could be resource intensive. One can reasonably argue that no single model can fully describe the growth of transportation networks for the complexities and multitude of dimensions it involves. Rather, this book approaches this complex topic from different angles with a series of theoretical, empirical, and simulation models. Though each model has its strengths and limitations, they together provide a comprehensive examination of transportation development.

The original contributions of this book are highlighted as follows:

- First, this book has demonstrated, with solid theoretical and empirical evidence, that the evolution of transportation systems is played out as the outcome of decentralized, interdependent decisions made by travelers, property owners, developers, financiers, and policy-makers at different levels from their respective initiatives. This highly original bottom-up, evolutionary view complements the traditional planning approach of metropolitan planning organizations (MPOs), which focuses on a centralized, top-down decision-making process by which authorities impose their investment decisions on transportation networks. Introducing agent-based simulation, this book has further demonstrated how simplistic, decentralized investment decisions could translate into a coherent network of facilities that exhibit spontaneous organization of collective attributes such as hierarchies, topologies, and sequences. In contrast with complicated, all-encompassing models that lack an explicit perspective, these models enable the display, capture, and analysis of these important network features throughout the temporal course of transport development, which has important implications for the calibration and validation of network growth models.
- Second, using the historical data of transportation networks, this book has validated the scientific proposition that the incremental deployment of transportation networks follows a logical path by which the accessibility to desirable land use activities is maximized. Based on the empirical findings, this book has introduced a series of *ex post* network growth models which use accessibility as the central organizing concept of the component models to explain travel behavior, pricing and investment policies, and location decisions of businesses and residences. Standard planning and engineering practices, concentrated on mobility improvement such as reducing total hours of delay or number of cars caught in congestion, are highly limited because they ignore people's motivations for travel. By focusing on accessibility rather than simple mobility measures, this book provides a more complete and meaningful picture of transportation and its role in our lives.
- Third, inspired by the idea of "greedy algorithm", this book has revealed that simulation developed based on decentralized local optimum decisions (such as "strongest-link" or "weakest-link" heuristics) was able to reproduce the observed

sequence of link additions/abandonment to a significant extent. This important finding supports to the path-dependent nature of transportation development which, while widely recognized in research, has not been previously corroborated. This finding also provides the foundation for the analytical representation of network growth through network degeneration and incremental connections in this book.

- Fourth, this book has analyzed the spillover effect between geographically adjacent jurisdictions and disclosed its reciprocal effects on the provision of a transportation network across the jurisdictions. Tradeoffs between the benefits and costs arising from network externalities and natural monopoly are found to be the key to understand the choice of spending structures and investment decisions on the cross-jurisdictional network.
- Fifth, this book has made methodological advances in network growth studies. It has invented measurements that are instrumental to capture the hierarchical and topological attributes of networks at a collective level, and to demonstrate the feasibility of abstract models of network growth to reveal these basic properties. This book also pioneers the use of non-parametric statistical analyses in the study of the sequential deployment of transportation networks.

Findings from throughout the book suggest that, the growth of transportation networks can be best described as an evolutionary process that features supply-demand interactions, independent players, and suboptimal (at best local optimal) decision-making subject to temporal and spatial constraints such as incomplete information, path dependence, spatial monopoly, network externalities, and interdependent economic / regulatory initiatives. The intrinsic characteristics of each transportation system such as its geography, demographics, economy, technology, culture, and politics add additional intricacies to this process. Out of all the complexities, this book manages to demonstrate with a wide spectrum of evidence that network growth is following a path that is not only logical in retrospect, but also predictable and manageable from a planning perspective. While independent players come into play under their respective interests, the next chapter will discuss if these interests could be better reconciled in an innovative planning process in the pursuit of more efficient and sustainable development of transportation systems for the overall welfare of society.

Chapter 17
Prospect

17.1 Perspectives on innovative planning

From the 1950s through the 1980s, the US federal government played a significant promotional role in the development of the Interstate system with large-scale capital investments. Now, a well-worn litany of seemingly intractable problems; congestion, environment, energy, land use, funding, and public health lead planners to suggest that there is widespread dissatisfaction with the institutions and political structures that are supposed to manage transportation. Well-understand (at least within the field) solutions such as road pricing remain stuck because either the problem is insufficiently severe or the institutions which are legally charged with the authority to deliver these solutions are insufficiently robust.

Transportation professionals who try to follow the traditional decision-making processes pioneered with the Interstate highway system, by which plans and designs are imposed top-down would find themselves in a difficult position in coping with innovations such as public-private partnerships and cross-jurisdictional congestion pricing. A re-consideration of conventional planning practice is in order. In response an evolutionary approach of transportation planning has seen increasing interest among both scholars and practitioners. The purpose of this book is certainly more than a review of transportation history or theoretic exploration of network modeling. Rather it contributes to the bigger scheme of innovative transportation planning. The prospects for the future success of evolutionary transportation planning depend on several factors, where we can see the contributions and implications this book has made.

An evolutionary approach to transportation planning must account for the evolutionary nature of transportation systems. An *evolutionary* decision-making process must be aware of the effects of its decisions on land use, environment, and urban development, as well as their interdependency with subsequent decisions. This approach renders decisions that address contemporary needs, but also explicitly considers how these decisions could play into the future in a favorable way. In spite of the extensive efforts that have been put into travel demand and land use forecasting, it is not clear how far these resource-intensive models can go in terms of assisting with transportation planning within time and budget constraints. The models of network growth, on the other hand, provide, in one sense, a sketch-level, but also a more engaged investigation into the pivotal determining factors such as accessibility and institutional arrangements in this evolutionary process, and could well serve the purpose of preliminary planing practice.

Evolutionary transportation planning also needs to explicitly consider the interdependent nature of current transportation systems. Regional planning organizations have been established largely in response to this need, but their current framework has been developed based on more of a jurisdictional consensus than a rational choice. For example, the conventional rule of "one member one vote" that obviously ignores the spatial heterogeneity of regional transportation systems is more like an outcome of political bargaining. How can an innovative approach perform better in directing the development of an interdependent transportation system? It first requires an explicit recognition of the prominent role that interdependent and/or conflicting interests have played in transportation systems, which we have demonstrated in the private promotion of Indiana interurbans, and in the legislative decision making processes of governmental road provision. More importantly, it requires explicit definitions of the scope and extent of individual players' needs in objective functions such that they could be accounted for in evolutionary transportation planning network system under alternative pricing or investment regimes.

Last but not the least, an evolutionary approach requires a sustainable vision to succeed, which involves an explicit treatment of externalities in transportation systems both on spatial and temporal dimensions. This essentially requires a redefinition of decision-making such that decision-makers decide not only to serve their own interests, but also align those interests with others. This applies both to individual users of transportation systems, but also to policy-makers, planners, and engineers who plan and deliver these systems. How could the incentive mechanism be re-designed in an innovative system of transportation planning to reconcile the interests of individual players such that they collectively do good for the society, the environment, and the future? Central to this effort are regulatory and pricing mechanisms of various forms by which externalities could be priced and internalized. The normative studies in this book, for example, have made the attempts to answer this question by testing whether the structure of governance can be rationally chosen to better coordinate conflicting local interests and maximize system welfare. Marginal cost pricing and cross-jurisdictional tolls are also examined in the attempt to address congestion and "free-rider" issues by charging road-users driving on congested roads or outside residence jurisdictions. Emissions trading (or "cap and trade") provides another example that a central authority can regulate the emission behaviors of local groups by establishing a community property right on clean air, and thus leading to a price on pollution. Extending these ideas, planners and decision-makers may ask themselves questions such as if a central authority can pose a "toll" or subsidy on individual jurisdictions or companies to better meet their interdependent interests arising from the positive/negative network externalities such as spatial spillovers and monopolies; how policy-makers themselves could be regulated so that the consequence of their decisions playing into the future could be taken into consideration in the present; how independent players of a system will react to such regulations and play into a new equilibrium? These are all open questions on which new insights are required to fulfill the strategic goals of transportation and minimize its negative consequences.

17.2 Future research

Aiming to capture the central features of the evolutionary transportation development process, we opted to sacrifice some important considerations in order to focus the investigation on the critical factors. This approach demonstrates how and why transportation networks evolve while leaving much room for improvement in the future. Extensions to this research could span across several directions.

With increasingly limited sources of funding, pricing and privatization of publicly-owned transportation networks have gained popularity in recent years. From an evolutionary perspective, the question arises whether the choice between the alternatives of taxes and tolls is a case-specific decision posed by a central authority, a natural result of system development, or something in-between. The game-theoretical models in this research have established the framework in which interdependent decisions made by individual residents, local governments, and a central government could be analyzed in a holistic process. Including the missing ingredient of tolls could enable these models to consider a wider spectrum of regulatory and pricing alternatives in the examination of the integrated decision-making processes.

The models herein treat public versus private, and centralized versus decentralized ownership structures of transportation systems separately. The reality, however, more likely sees a mixed structure with intertwined ownership. While mixed ownership of transportation systems has been examined in the literature using empirical data, little has been done at a theoretical level in modeling or analyzing the decision-making processes under a mixed ownership regime. An extension of our analytical or simulation models to include the organization of public-private partnerships or highway fund allocation would be useful.

From a modeling perspective, an explicit consideration of the two-way transportation / land use interaction would be a valuable extension to the current network growth models that treat land use as exogenous. In contrast with those comprehensive but expensive approaches such as hedonic analysis and integrated transportation-land use simulation, we could follow the logic of this book and focus mainly on the critical driving factors of the integrated process from an evolutionary view. Chapter 12 has touched upon this issue by demonstrating based on simple decision rules the spontaneous organization of a transportation network coupled with the relocation of businesses and residences surrounding the network. This effort could be extended in the future towards a more realistic representation of the co-evolutionary process of network and land use. The risk is the essence of the problem is lost with the additional details required for realism.

In addition, a convincing procedure of calibration and validation for network growth models would be important for their application in planning practice. Most current transportation planning models are only calibrated partially and separately for some of its component models. Chapter 4 suggests a new way to validate network growth models against the observed sequence of link additions or graph-theoretical measures of observed network features at a collective level. Chapter 9 applied these methods to validate the model of network degeneration using the Indiana interurban network during its decline phase. Despite these efforts, validation of a network growth model remains a challenging task given its multitude and magnitude of complexities, and deserves further exploration.

Appendix A
Mathematical Derivations of Governance Choice Analysis

A.1 Pigouvian governments model

On a one-dimension space, it is easy to check the travel times that an individual resident living at x needs to reach the centers of each district:

To reach the center of residence district,

$$t(x,i) = \begin{cases} \frac{-x}{v_i} & -L/2 \leq x \leq 0 \\ \frac{x}{v_i} & L/2 \leq x \leq L \end{cases}$$

To reach the center of neighboring district,

$$t(x,j) = \frac{L/2 - x}{v_i} + \frac{L/2}{v_j}$$

With the assumption of fixed demand, the total travel time an individual resident living at x spends on roads for a planning period can be calculated as:

$$t_x = \sum_i \left(2\rho(x,i)t(x,i)\right)$$
$$= \begin{cases} 2k\eta \frac{-x}{v_i} + 2k(1-\eta)(\frac{L/2-x}{v_i} + \frac{L/2}{v_j}) & -L/2 \leq x \leq 0 \\ 2k\eta \frac{x}{v_i} + 2k(1-\eta)(\frac{L/2-x}{v_i} + \frac{L/2}{v_j}) & L/2 \leq x \leq L \end{cases}$$

Assuming a uniform distribution of residents across the space, we obtained the total travel time residents in district i spend on roads in the district by Integrating individual travel time across a district:

$$T_i = \int_{-L/2}^{L/2} r_x t_x dx$$

$$= \int_{-L/2}^{0} \left\{ 2rk\eta \tfrac{-x}{v_i} + 2rk(1-\eta)(\tfrac{L/2-x}{v_i} + \tfrac{L/2}{v_j}) \right\} dx$$

$$+ \int_{0}^{L/2} \left\{ 2rk\eta \tfrac{x}{v_i} + 2rk(1-\eta)(\tfrac{L/2-x}{v_i} + \tfrac{L/2}{v_j}) \right\} dx$$

$$= \frac{(1-\eta/2)rkL^2}{v_i} + \frac{(1-\eta)rkL^2}{v_j}$$

A benevolent local government considers both resources cost (road spending) and user cost (travel time cost) during decision making of road speed:

$$U_i^D = -C_i(v_0, v_i) - \tau T_i$$
$$= -\tau \alpha_0 v_0^{\alpha_1} \alpha_2 (v_i - v_0) L^{1+\alpha_3} - \frac{(1-\eta/2)rkL^2}{v_i} - \frac{(1-\eta)rkL^2}{v_j}$$

The road speed chosen by a local government is obtained by taking the first derivative of its utility function with respect to the road speed in this district,

$$\frac{\partial U_i^D}{\partial v_i^D} = 0$$

$$\Rightarrow -\frac{(1-\eta/2)rkL^2}{\left(v_i^D\right)^2} + \tau \alpha_0 v_0^{\alpha_1} \alpha_2 L^{1+\alpha_3} = 0$$

$$\Rightarrow v_i^D = \sqrt{\frac{(1-\eta/2)rkL^{1-\alpha_3}}{\tau \alpha_0 v_0^{\alpha_1} \alpha_2}}$$

Under centralized decision making, it is assumed the central government aims to maximize the aggregate surplus of both districts while putting different weights on them,

$$U^C = \beta_i U_i^C + \beta_j U_j^C$$
$$= -\beta_i \left\{ \frac{(1-\eta/2)rkL^2}{v_i} + \frac{(1-\eta)rkL^2}{v_j} + \tau \alpha_0 v_0^{\alpha_1} \alpha_2 \frac{(v_i-v_0+v_j-v_0)}{2} L^{1+\alpha_3} \right\}$$
$$-(1-\beta_i) \left\{ \frac{(1-\eta/2)rkL^2}{v_j} + \frac{(1-\eta)rkL^2}{v_i} + \tau \alpha_0 v_0^{\alpha_1} \alpha_2 \frac{(v_i-v_0+v_j-v_0)}{2} L^{1+\alpha_3} \right\}$$

Road speeds chosen for two districts can be obtained by taking the first derivative of the central government's utility function with respect to the road speed in the two districts, respectively:

$$\frac{\partial U^C}{\partial v_i^C} = 0$$

$$\Rightarrow -\frac{\beta_i(1-\eta/2)rkL^2}{\left(v_i^C\right)^2} + \tfrac{1}{2}\tau \alpha_0 v_0^{\alpha_1} \alpha_2 L^{1+\alpha_3} - \frac{(1-\beta_i)(1-\eta)rkL^2}{\left(v_i^C\right)^2} = 0$$

$$\Rightarrow v_i^C = \sqrt{\frac{(2-(2-\beta_i)\eta)rkL^{1-\alpha_3}}{\tau \alpha_0 v_0^{\alpha_1} \alpha_2}}$$

Without loss of generality, it is assumed that the spending policies at a central level is skewed against district i (i.e., $0 \le \beta_i \le 0.5$). District j will benefit from the misallocation at the expense of district i. Therefore, if district i votes for centralization, district j will always vote for centralization. In this case, the adoption of a centralized structure depends on the following requirement:

$$U_i^D <= U_i^C$$
$$\Leftrightarrow -U_i^C <= -U_i^D$$
$$\Leftrightarrow \frac{(1-\eta/2)rkL^2}{v_i^C} + \tau\alpha_0 v_0^{\alpha_1} \alpha_2 \frac{v_i^C+v_j^C-2v_0}{2}L^{1+\alpha_3} + \frac{(1-\eta)rkL^2}{v_j^C}$$
$$\leq \frac{(1-\eta/2)rkL^2}{v_i^D} + \frac{(1-\eta)rkL^2}{v_j^D} + \tau\alpha_0 v_0^{\alpha_1} \alpha_2 (v_i^D - v_0)L^{1+\alpha_3}$$
$$\Leftrightarrow g(\eta,\beta_i) =$$
$$(1-\tfrac{\eta}{2})\left(\sqrt{\frac{2-(2-\beta_i)\eta}{4(1-\eta/2)}} + \sqrt{\frac{2-(1+\beta_i)\eta}{4(1-\eta/2)}} + \sqrt{\frac{1-\eta/2}{2-(2-\beta_i)\eta}} - 2\right) - (1-\eta)\left(1 - \sqrt{\frac{1-\eta/2}{2-(1+\beta_i)\eta}}\right) \leq 0$$

A.2 Citizen-candidate model

According to the assumption that uniform taxes are charged to finance road provision under either centralized or decentralized decision making, the taxes that a resident needs to pay can be calculated as:

Under decentralized decision making,

$$f^D = \frac{\alpha_0\alpha_2 v_0^{\alpha_1}(v_i - v_0)L^{1+\alpha_3}}{rL}$$

Under centralized decision making,

$$f^C = \frac{\alpha_0\alpha_2 v_0^{\alpha_1}(v_i + v_j - 2v_0)L^{1+\alpha_3}}{2rL}$$

Under decentralized decision making, residents vote for a road speed that maximizes their individual utility. The utility function of a resident takes the following form:

$$u_{x,i}^D = -\tau t_x - f^D$$
$$= \begin{cases} -2k\tau\left(-\eta\frac{1}{v_i} + (1-\eta)\frac{\frac{L}{2}-x}{v_i} + (1-\eta)\frac{\frac{L}{2}}{v_j}\right) - \frac{1}{r}\alpha_0\alpha_2 v_0^{\alpha_1}(v_i - v_0)L^{\alpha_3} & -L/2 \leq x \leq 0 \\[2ex] -2k\tau\left(\eta\frac{1}{v_i} + (1-\eta)\frac{\frac{L}{2}-x}{v_i} + (1-\eta)\frac{\frac{L}{2}}{v_j}\right) - \frac{1}{r}\alpha_0\alpha_2 v_0^{\alpha_1}(v_i - v_0)L^{\alpha_3} & L/2 \leq x \leq L \end{cases}$$

The favored speed of an individual resident can thus be derived by taking the first derivative of his/her utility function with respect to the road speed in the residence district:

$$v_{x,i}^{D} = \begin{cases} \sqrt{\dfrac{(1-\eta-\frac{2x}{L})\tau rkL^{1-\alpha_3}}{\alpha_0 v_0^{\alpha_1}\alpha_2}} & -L/2 \leq x \leq 0 \\[4ex] \sqrt{\dfrac{(1-\eta+\frac{2x}{L}(2\eta-1))\tau rkL^{1-\alpha_3}}{\alpha_0 v_0^{\alpha_1}\alpha_2}} & L/2 \leq x \leq L \end{cases}$$

Note that whether a resident lives to the left or right of the district center, the preferred road speed increases monotonically with the distance from residential location to the center. Suppose residents living at x_- on the half farther from the border between two districts and at $x+$ on the half closer to the border prefer the same level of road spending, it is easy to check that:

$$\frac{-2x_-}{L} = \frac{2x_+}{L}(2\eta - 1)$$

In order to become a median voter, the median voter theorem suggests there should half of the voters prefer a road speed that is lower than the median-type road speed. Based on the assumption of uniform distribution of residents across space, a median voter lives in a location that satisfies:

$$\frac{x_-}{-\frac{L}{2}} + \frac{x_+}{\frac{L}{2}} = \frac{1}{2}$$

Solving the above two equations jointly, we obtain the locations where median voters live:

$$\begin{cases} x_+^M = \frac{1}{8\eta}L \\[1ex] x_-^M = \frac{1-2\eta}{8\eta}L \end{cases}$$

Accordingly, the elected representative commits a road speed preferred by median voters as:

$$v_i^D = \sqrt{\frac{(1-\eta+\frac{2\eta-1}{4\eta})\tau rkL^{1-\alpha_3}}{\alpha_0 v_0^{\alpha_1}\alpha_2}}$$

Under centralized decision making, it is assumed that representatives elected from districts will form minimum winning coalitions in the legislature. Each coalition has equal possibility to attract majority of the legislature.

Suppose the representative from district i is selected to determine road spending policy, it is assumed the representative aims to choose spending levels for both districts while aiming to maximize the utility of the median voter in home district, which can be translated into:

$$\begin{aligned} u_{x_-^M,i}^{C} &= -\tau t_x - f^C \\ &= -\tau\left(2k\eta\frac{-x_-^M}{v_i} + 2k(1-\eta)(\frac{L/2-x_-^M}{v_i} + \frac{L/2}{v_j})\right) - \frac{\alpha_0\alpha_2 v_0^{\alpha_1}(v_i+v_j-v_0)L^{1+\alpha_3}}{2rL} \\ &= -2k\tau\left(-\eta\frac{1}{v_i} + (1-\eta)\frac{L/2-x_-^M}{v_i} + (1-\eta)\frac{L/2}{v_j}\right) - \frac{1}{2r}\alpha_0\alpha_2 v_0^{\alpha_1}(v_i+v_j-v_0)L^{\alpha_3} \end{aligned}$$

In this case, it is easy to derive the levels of road spending for both districts:

$$\begin{cases} v_i^C = \sqrt{\dfrac{2(1-\eta+\frac{2\eta-1}{4\eta})\tau rkL^{1-\alpha_3}}{\alpha_0 v_0^{\alpha_1}\alpha_2}} \\ v_j^C = \sqrt{\dfrac{2(1-\eta)\tau rkL^{1-\alpha_3}}{\alpha_0 v_0^{\alpha_1}\alpha_2}} \end{cases}$$

Thus, the aggregate welfare in district i can be calculated as:

$$U_i^C = -\frac{1}{2}\left(\frac{(1-\eta/2)\tau rkL^2}{v_i^C} + \alpha_0 v_0^{\alpha_1}\alpha_2(\frac{v_i^C+v_j^C}{2}-v_0)L^{1+\alpha_3} + \frac{(1-\eta)\tau rkL^2}{v_j^C}\right)$$

Similarly, if the representative from district j wins the majority, the levels of road spending for both districts can be derived as:

$$U_i^C = -\frac{1}{2}\left(\frac{(1-\eta/2)\tau rkL^2}{v_j^C} + \alpha_0 v_0^{\alpha_1}\alpha_2(\frac{v_i^C+v_j^C}{2}-v_0)L^{1+\alpha_3} + \frac{(1-\eta)\tau rkL^2}{v_i^C}\right)$$

As can be seen in either case, the policy maker tends to oversupply the road infrastructure in his/her home district, and undersupply that in the neighboring district.

There is equal probability for the representative from district i and that from district j to take the decision power in the legislature. Thus the expected utility function of an elected representative from district i can be formulated as:

$$U_i^C = -\frac{1}{2}\left(\frac{(1-\eta/2)\tau rkL^2}{v_i^C} + \alpha_0 v_0^{\alpha_1}\alpha_2(\frac{v_i^C+v_j^C}{2}-v_0)L^{1+\alpha_3} + \frac{(1-\eta)\tau rkL^2}{v_j^C}\right)$$
$$-\frac{1}{2}\left(\frac{(1-\eta/2)\tau rkL^2}{v_j^C} + \alpha_0 v_0^{\alpha_1}\alpha_2(\frac{v_i^C+v_j^C}{2}-v_0)L^{1+\alpha_3} + \frac{(1-\eta)\tau rkL^2}{v_i^C}\right)$$

Policy centralization is adopted when both districts vote for centralization, thus the condition on which centralization is chosen can be derived as:

$$U_i^D <= U_i^C$$
$$\Leftrightarrow -U_i^C <= -U_i^D$$
$$\Leftrightarrow \frac{1}{2}\left(\frac{(1-\eta/2)\tau rkL^2}{v_i^C} + \alpha_0 v_0^{\alpha_1}\alpha_2(\frac{v_i^C+v_j^C}{2}-v_0)L^{1+\alpha_3} + \frac{(1-\eta)\tau rkL^2}{v_j^C}\right)$$
$$+\frac{1}{2}\left(\frac{(1-\eta/2)\tau rkL^2}{v_j^C} + \alpha_0 v_0^{\alpha_1}\alpha_2(\frac{v_i^C+v_j^C}{2}-v_0)L^{1+\alpha_3} + \frac{(1-\eta)\tau rkL^2}{v_i^C}\right)$$
$$\leq \frac{(1-\eta/2)\tau rkL^2}{v_i^D} + \frac{(1-\eta)\tau rkL^2}{v_i^D} + \alpha_0 v_0^{\alpha_1}\alpha_2(v_i^D-v_0)L^{1+\alpha_3}$$
$$\Leftrightarrow h(\eta) = \frac{(\sqrt{2}/2-2)(1-3\eta/4)}{\sqrt{1-\eta+\frac{2\eta-1}{4\eta}}} + (\sqrt{2}/2-1)\sqrt{1-\eta+\frac{2\eta-1}{4\eta}} + \frac{2-7\eta/4}{\sqrt{2(1-\eta)}} <= 0$$

References

Adamatzky, A. and Jones, J. (2009), 'Road Planning with Slime Mould: If Physarum Built Motorways It Would Route M6/M74 through Newcastle', Int J Bifurcation and Chaos (In Press).

Adams, J. and VanDrasek, J. (1993), *Minneapolis-St. Paul: People, Place, and Public Life*, University of Minnesota Press.

Ahlfeldt, G. M. and Wendland, N. (2010), 'Fifty Years of Urban Accessibility : The Impact of Urban Railway Network on the Land Gradient in Industrializing Berlin', Regional Science and Urban Economics (forthcoming).

Airports Council International (2006), Traffic Movements 2005, Technical report, Airports Council International.
URL: *http://www.aci.aero/*

Albert, R., Jeong, H. and Barabasi, A. L. (1999), "Diameter of The World-wide Web", *Nature* , Vol. 401, pp. 130–131.

Alberti, M. and Waddell, P. (2000), "An Integrated Urban Development and Ecological Simulation Model", *Integrated Assessment* , Vol. 1, pp. 215–227.

Alperovich, G., Kemp, M. and Goodman, K. (1977), 'An Econometric Model of Bus Transit Demand and Supply', *The Urban Institute Working Paper No. 5032-1-4, Washington, D.C.*

Anderson, D. and Mohring, H. (1997), Congestion Costs and Congestion Pricing, *in* D. Greene, D. Jones and M. Delucchi, eds, 'The Full Costs and Benefits of Transportation: Contributions, Theory and Measurement', New York: Springer.

Arthur, W. B. (1994), *Increasing Returns and Path Dependence in the Economy*, Ann Arbor, Michigan: University of Michigan Press.

Atack, J. and Margo, R. (2011), 'The Impact of Access to Rail Transportation on Agricultural Improvement: The American Midwest as a Test Case, 1850-1860', Journal of Transportation and Land Use (forthcoming).

Balch, T. (2000), "Hierarchic Social Entropy: an Information Theoretic Measure of Robot Team Diversity", *Autonomous Robots* , Vol. 8, p. 209237.

Barabási, A. (2002), *Linked: the New Science of Networks*, Perseus Publication.

Barabási, A. and Bonabeau, E. (2003), "Scale-Free Networks", *Scientific American* , Vol. 288.

Barabási, A. L. and Albert, R. (1999), "Emergence of Scaling in Random Networks", *Science* , Vol. 286, pp. 509–512.

Baran, P. (1964), On Distributed Communications, Memorandum RM-3764-PR, The RAND Corporation.

Barrat, A., Barthélemy, M. and Vespignani, A. (2004), "Modeling the Evolution of Weighted Networks", *Physical Review E* , Vol. 70, APS, pp. 66–149.

Barthélemy, M. (2010), Spatial Networks.

Barthélemy, M. and Flammini, A. (2006), "Optimal Traffic Networks", *Journal of Statistical Mechanics* , p. L07002.

Barthélemy, M. and Flammini, A. (2009), "Co-evolution of Density and Topology in a Simple Model of City Formation", *Networks and Spatial Economics* , Vol. 9(3), pp. 401–425.

Bates, J., Brewer, M., Hanson, P., McDonald, D. and Simmonds, D. (1991), Building a Strategic Model for Edinburgh, *in* 'Proceedings of Seminar D, PTRC 19th Summer Annual Meeting. PTRC, London'.

Batty, M. and Xie, Y. (1994), "From Cells to Cities", *Environment and Planning B* , Vol. 21, pp. 31–48.

Bertaud, A. (2001), A Measure of the Spatial Organization of 7 Large Cities.
   URL: *http://alain-bertaud.com/images/AB_Metropolis_Spatial_Organization.pdf*

Bertolini, L. (2007), "Evolutionary Urban Transportation Planning: An Exploration", *Environment and Planning A* , Vol. 39, pp. 1998–2019.

Besley, T. and Coate, S. (2003), "Central versus Local Provision of Public Goods: a Political Economy Analysis", *Journal of Public Economics* , Vol. 87(4), pp. 2611–2637.

Black, W. (1971), "An Iterative Model for Generating Transportation Networks", *Geographical Analysis* , Vol. 3, pp. 283–288.

Blumenfeld-Lieberthal, E. (2009), "The Topology of Transportation Networks: A Comparison between Different Economies", *Networks and Spatial Economics* , Vol. 9(3), pp. 427–458.

Bogart, D. (2005), "Turnpike Trusts, Infrastructure Investment, and the Road Transportation Revolution in Eighteenth-Century England", *The Journal of Economic History* , Vol. 65, pp. 540–543.

Bogart, D. (2009), "Inter-modal Network Externalities and Transport Development: Evidence from Roads, Canals, and Ports during the English Industrial Revolution", *Networks and Spatial Economics* , Vol. 9(3), pp. 309–338.

Borger, B. D., Proost, S. and Dender, K. V. (2005), "Congestion and Tax Competition in a Parallel Network", *European Economic Review* , pp. 2013 – 2040.

Borley, H. (1982), *Chronology of London Railways*, Railway and Canal Historical Society, Oakham, Leicester, England.

Boyce, D. (2007), An Account of a Road Network Design Method: Experssway Spacing, System Configuration and Economic Evaluation, *in* 'Infrastructure Problems under Population Decline', Berlin: Bediner Wissenschafts-Vedag, pp. 1–30.

Buchanan, J. and Tullock, G. (1962), *The Calculus of Consent, Logical foundations of Constitutional Democracy*, Ann Arbor, The University of Michigan Press.

Bureau of Public Roads (1964), *Traffic Assignment Manual*, U.S. Dept. of Commerce, Urban Planning Division, Washington D.C.

Business Travel News Online (2004), 'Siegel: Philadelphia Could Be US Airways' Last Stand'.
URL: *http://www.businesstravelnews.com/More-News/Articles/Siegel–Philadelphia-Could-Be-US-Airways–Last-Stand/*

Button, K. (1998), "Infrastructure Investment, Endogenous Growth and Economic Convergence", *Annals of Regional Science* , Vol. 32, pp. 145–162.

Byers, J. P. (1998), Breaking the Ground Plane: The Evolution of Grade Separated Cities in North America, PhD thesis, University of Minnesota, Department of Geography.

Casson, M. (2009), "The Efficiency of the Victorian British Railway Network: A Counterfactual Analysis", *Networks and Spatial Economics* , Vol. 9(3), pp. 339–378.

Center for Economic and Social Inclusion (2006), National Statistics: First Release: Labour Market Statistics, Technical report, Center for Economic and Social Inclusion, London.
URL: *http://www.cesi.org.uk/statsdocs/0602/lmslond0206.pdf Table 12*

Cervero, R. and Hansen, M. (2002), "Induced Travel Demand and Induced Road Investment: A Simultaneous Equation Analysis", *Journal of Transport Economics and Policy* , Vol. 36, LSE and the University of Bath, pp. 469–490.

Chachra, V., Ghare, P. and Moore, J. (1979), *Applications of Graph Theory Algorithms*, North Holland, New York.

Chatterjee, S. (2003), "Agglomeration Economies: The Spark That Ignites a City? ", *Business Review* , Vol. 4, pp. 6–13.

Cheikbossian, G. (2004), "Lobbying and Rent-seeking for Public Goods in a Fiscally Centralized System", *Annales D Economie Et De Statistique* , pp. 331–351.

Chen, W. (2004), Highway Network Evolution Models, Master's thesis, University of Minnesota at Twin Cities.

Christaller, W. (1933), *Die zentralen Orte in Suddeutschland*, Jena: Gustav Fischer.

Collie, M. (1988), "The Legislature and Distributive Policy Making in Formal Perspective", *Legislative Studies Quarterly* , Vol. 13, pp. 427–458.

Corbett, M., Xie, F. and Levinson, D. (2009), "Evolution of the Second-Story City: The Minneapolis Skyway System", *Environment and Planning B* , Vol. 36(4), pp. 711–724.

Courtat, T., Gloaguen, C. and Douady, S. (2010), Mathematics and Morphogenesis of the City, A Geometrical Approach.

Csányi, G. and Szendröi, B. (2004), "Fractal-small-world Dichotomy in Real-world Networks", *Physical Review E* , Vol. 70.

Curry, L. (1964), "The Random Spatial Economy: An Exploration in Settlement Theory", *Annual of the Association of Merican Geographers* , Vol. 54, pp. 138–146.

Davis, G. and Sanderson, K. (2002), Building Our Way Out Of Congestion, Final Report MN/RC-2002-01, Minnesota Department of Transportation.

De Montis, A., Barthelemy, M., Chessa, A. and Vespignani, A. (2006), The Structure of Inter-Urban Traffic: A Weighted Network Analysis.
URL: *http://arxiv.org/abs/physics/0507106*

de Palma, A. and Lindsey, R. (2000), "Private Toll Roads: Competition under Various Ownership Regimes", *Annals of Regional Science* , Vol. 34, pp. 13–35.

de Solla Price, D. J. (1965), "Networks of Scientific Papers", *Science* , Vol. 149, pp. 510–515.

Demange, G. and Wooders, M. (2005), *Group Formation in Economics: Networks, Clubs and Coalitions*, Cambridge University Press.

Derrible, S. and Kennedy, C. (2010), "The Complexity and Robustness of Metro Networks", *Physica A: Statistical Mechanics and its Applications* , Vol. 389, pp. 3678–3691.

Diers, J. and Isaacs, A. (2006), *Twin Cities by Trolley: The Streetcar Era in Minneapolis and St. Paul*, University of Minnesota Press.

Dorogovtsev, S. and Mendes, J. (2002), "Evolution of Networks", *Advance in Physics* , Vol. 51, p. 1079.

Downs, A. (1957), *An Economic Theory of Democracy*, New York: Harper and Row.

Downs, A. (1992), *Stuck in Traffic: Coping with Peak-Hour Traffic Congestion*, Brookings Institution Press.

Dur, R. and Roelfsema, H. (2005), "Why does Centralisation Fail to Internalise Policy Externalities?", *Public Choice* , Vol. 122, p. 395416.

Economides, N. (1996), "The Economics of Networks", *International Journal of Industrial Organization* , Vol. 14(6), pp. 673–699.

El-Geneidy, A., Kastelberger, L. and Abdelhamid, H. T. (2011), 'Montréal's Roots: Exploring the Growth of Montréal's Indoor City', Journal of Transportation and Land Use (forthcoming).
URL: *http://tram.mcgill.ca/Research/Publications/Montreal_indoor_city_final_in_JTLU.pdf*

El-Geneidy, A. M. and Levinson, D. (2006), Access to Destinations: Development of Accessibility Measures, Technical report, Center for Transportation Research, University of Minnesota.

Epple, D. and Nechyba, T. (2004), Fiscal Decentralization, *in* V. Henderson and J. Thisse, eds, 'Handbook of Regional and Urban Economics', Vol. 4, North Holland, Elsevier, pp. 2423–2480.

Erath, A., Löchl, M. and Axhausen, K. W. (2009), "Graph-theoretical Analysis of the Swiss Road Network over Time", *Networks and Spatial Economics* , Vol. 9(3), pp. 379–400.

Eugster, M. J. A. (2010), 'H&M Travelling Salesman Problem', Retrieved November 7, 2010.
URL: *http://www.statistik.lmu.de/ eugster/teaching/rsi/graphs-tsp-hm.html*

Ewing, R. (2000), "Sketch Planning a Street Network ", *Transportation Research Record* , Vol. 1722, pp. 75–79.

Federal Highway Administration (2002), 'Highway Statistics 2002'.
URL: *http://www.fhwa.dot.gov/policy/ohim/hs02/hf.htm*

Fullerton, B. (1975), *The Development of British Transport Networks*, Oxford University Press.

Fulton, L., Meszler, R. and Thomas, J. (2000), "A Statistical Analysis of Induced Travel Effects in the US Mid-Atlantic Region", *Journal of Transportation and Statistics* , Vol. 3, pp. 1–14.

Garrison, W. L. (1960), "Connectivity of the Interstate Highway System", *Regional Science Association, Papers and Proceedings* , Vol. 6, pp. 121–137.

Garrison, W. L. and Marble, D. F. (1962), The Structure of Transportation Networks, Technical report.

Garrison, W. and Levinson, D. (2006), *The Transportation Experience: Policy, Planning, and Deployment*, Oxford University Press, USA.

Gastner, M. T. and Newman, M. E. J. (2006), "The Spatial Structure of Networks", *The European Physical Journal B* , Vol. 49, pp. 247–252.

Gaudry, M. (1975), "An Aggregate Time-series Analysis of Urban Transit Demand: the Montreal Case", *Transportation Research* , Vol. 9, p. 249258.

Gibbons, R. (1992), *Game Theory for Applied Economists*, Princeton University Press.

Gómez-Ibáñez, J., Tye, W. B. and Winston, C. (1999), *Essays in Transportation Economics and Policy*, Washington, D.C.: The Brookings Institution.

Granger, C. W. J. (1969), "Investigating Causal Relations by Econometric Models and Cross-spectral Methods", *Econometrica* , Vol. 37, pp. 424–438.

Green, H. and Truesdell, L. (1937), 'Census Tracts in American Cities: A Brief History of the Census Tract Movement, with an Outline of Procedure and Suggested Modifications'.
URL: *http://www2.census.gov/prod2/decennial/documents/21012490_TOC.pdf*

Güth, W., Schmittberger, R. and Schwarze, B. (1982), "An Experimental Analysis of Ultimatum Bargaining", *Journal of Economic Behavior and Organization* , Vol. 3(4), pp. 367–388.

Gutin, G. and Karapetyan, D. (2008), *Greedy Algorithms*, InTech, chapter Greedy Like Algorithms for the Traveling Salesman and Multidimensional Assignment Problems, pp. 291–304.

Guttenberg, A. (1960), "Urban Structure and Growth", *Journal of American Inst. Planners* , Vol. 26, pp. 104–110.

Haggett, P. (1966), *Locational Analysis in Human Geography*, New York: St. Martins Press.

Haggett, P. and Chorley, J. C. (1969), *Network Analysis in Geography*, Butler and Tanner Ltd.

Haley, R. M. (1936), The American Electric Railway Interurban, PhD thesis, Northwestern University.

Hansen, W. (1959), "How Accessibility Shapes Land Use", *Journal of American Inst. Planners* , Vol. 25, pp. 73–76.

Hanson, N. (2001), *The Dreadful Judgement: The True Story of the Great Fire of London*, New York: Doubleday.

Hanson, S., ed. (1995), *The Geography of Urban Transportation*, Guilford Publications.

Haynes, K. E., Gifford, J. L. and Pelletiere, D. (2005), "Sustainable Transportation Institutions and Regional Evolution: Global and Local Perspectives", *Journal of Transport Geography* , Vol. 13, p. 207221.

Helbing, D., Keltsch, J. and Molnr, P. (1997), "Modeling the Evolution of Human Trail Systems", *Nature* , Vol. 388, p. 47.

Henrich, J., Boyd, R., Bowles, S., Camerer, C., Fehr, E. and Gintis, H. (2004), *Foundations of Human Sociality: Economic Experiments and Ethnographic Evidence from Fifteen Small-Scale Societies*, Oxford University Press.

Higgins, J. J. (2003), *Introduction to Modern Nonparametric Statistics*, Duxbury Press.

Hilton, G. W. and Due, J. F. (1960), *The Electric Interurban Railways in America*, Stanford University Press, Stanford.

Hommels, A. (2005), *Unbuilding Cities: Obduracy in Urban Socio-technical Change*, Cambridge, Mass.: MIT Press.

Huang, Y. (2004), "Urban Spatial Patterns and Infrastructure in Beijing", *Land Lines* , Vol. 16, p. 4.

Huff, D. (1963), "A Probabilistic Analysis of Shopping Trade Areas", *Land Economics* , Vol. 39, pp. 81–90.

Humplick, F. and Moini-Araghi, A. (1996*a*), 'Decentralized Structures for Providing Roads: A Cross-Country Comparison', *Policy Research Working Paper 1658. World Bank, Policy Research Department, Washington, DC*.

Humplick, F. and Moini-Araghi, A. (1996*b*), 'Is There an Optimal Structure for Decentralized Provision of Roads?', *Policy Research Working Paper 1657. World Bank, Policy Research Department, Washington, DC*.

Hutchinson, B. (1974), *Principles of Urban Transportation Systems Planning*, New York: McGraw-Hill.

Iacono, M., Levinson, D. and El-Geneidy, A. (2008), "Models of Transportation and Land Use Change: A Guide to the Territory", *Journal of Planning Literature* , Vol. 22, pp. 323–340.

Institute of Transportation Engineers (1997), 'ITE Trip Generation, 6th Edition', Institute of Transportation Engineers, Washington D.C.

International Road Assessment Programme (2008), 'International Transport Statistics', http://www.iraptranstats.net/infra.

Isaacs, A. and Diers, J. (2007), *Twin City Lines: The Streetcar Era in Minneapolis and St. Paul*, University of Minnesota Press.

Jackson, M. and Wolinsky, A. (1996), "A Strategic Model of Social and Economic Networks", *Journal of Economic Theory* , Vol. 71, pp. 44–74.

Jacob, B. (1984), *Skyway Typology: A Study of the Minneapolis Skyways*, AIA Press, Washington, D.C.

Jardine, L. (2002), *On a Grander Scale: The Outstanding Career of Sir Christopher Wren*, HarperCollins.

Jeong, H., Gombor, B., Albert, R., Oltwai, Z. N. and Barabási, A. L. (2000), "The Large-scale Organization of Metabolic Networks", *Nature* , Vol. 407, pp. 651–654.

Jiang, B. (2005), Small World Modelling for Complex Geographic Environments, *in* J. Portugali, ed., 'Complex Artificial Environments', Springer, pp. 259–271.

Jiang, B. (2007), "A Topological Pattern of Urban Street Networks: Universality and Peculiarity", *Physica A: Statistical Mechanics and its Applications* , Vol. 384, pp. 647 – 655.

Jiang, B. and Claramunt, C. (2004), "Topological Analysis of Urban Street Networks", *Environment and Planning B* , Vol. 31, pp. 151–162.

Johnson, R. and Libecap, G. (2000), 'Political Processes and the Common Pool Problem: the Federal Highway Trust Fund', *Working Paper, Department of Economics. Tucson, AZ: University of Arizona.*

Kansky, K. (1963), 'Structure of Transportation Networks: Relationships between Network Geometry and Regional Characteristics', Chicago: Univ. of Chicago, Department of Geography, Research paper 84.

Kaufman, S. (1985), *The Skyway Cities*, CSPI, Minneapolis, Minnesota.

Kerin, R., Varadarajan, P. and Peterson, R. (1996), "First-Mover Advantage: A Synthesis, Conceptual Framework, and Research Propositions", *Journal of Marketing* , Vol. 56(4), p. 3352.

King, D. (2011), 'Developing Densely: Estimating the Effect of Subway Growth on New York City Land Uses', Journal of Transportation and Land Use (forthcoming).
URL: *http://www.civil.ist.utl.pt/wctr12_lisboa/WCTR_General/documents/02056.pdf*

King, L. (1985), *Central Place Theory*, SAGE Publication Ltd, London.

Kirk, R. S. (2004), 'Federal-Aid Highway Program: Donor-Donee State Issues', *CRS Report for Congress, Order Code RL31735.*

Knight, B. (2002), "Endogenous Federal Grants and Crowd-Out of State Government Spending: Theory and Evidence from the Federal Highway Aid Program", *American Economic Review* , Vol. 92(1), pp. 71–92.

Knight, B. (2003), 'Parochial Interests And The Centralized Provision Of Local Public Goods: Evidence From Congressional Voting On Transportation Projects', *National Bureau Of Economic Research, Working Paper No. W9748.*

Kolars, J. and Malin, H. J. (1970), "Population and Accessibility: An Analysis of Turkish Railroads", *The Geographical Review* , Vol. 60, pp. 229–246.

Kondratieff, N. (1987), *The Long Wave Cycle*, New York, NY: Richardson and Snyder.

Kramer, M. B. (2004), 'Minnesota Leasing Guide - First Quarterly Edition', Law Bulletin Publishing Company, Chicago, IL.

Krugman, P. R. (1996), *The Self-Organizing Economy*, Blackwell Publishers.

Lachene, R. (1965), "Networks and the Location of Economic Activities", *Regional Science Association, Papers* , Vol. 14, pp. 183–196.

Lam, L. (1995), "Active Walker Model Walker Models for Complex Systems", *Chaos, Solitons and Fractals* , Vol. 6, pp. 267–285.

Lam, L. and Pochy, R. (1993), "Active-walker Models: Growth and Form in Nonequilibrium Systems", *Computation simulation* , Vol. 7, p. 534.

Lämmer, S., Gehlsen, B. and Helbing, D. (2006), "Scaling Laws in the Spatial Structure of Urban Road Networks", *Physica A* , Vol. 363, pp. 89–95.

LeBlanc, L. J. (1975), "An Algorithm for the Discrete Network Design Problem", *Transportation Science* , Vol. 9(3), pp. 183–199.

Lee, D., J. (1973), "Requiem for Large-Scale Models", *Journal of the American Institute of Planners* , Vol. 40, pp. 163–178.

Leeming, J. J. (1969), *Road Accidents: Prevent or Punish*, Cassell.

Levinson, D. (2002), *Financing Transportation Networks*, Edward Elgar Publishing.

Levinson, D. (2008*a*), "Density and Dispersion: The Co-Development of Land Use and Rail in London", *Journal of Economic Geography* , Vol. 8(1), pp. 55–57.

Levinson, D. (2008*b*), "The Orderliness Hypothesis: Does Population Density Explain the Sequence of Rail Station Opening in London?", *Journal of Transport History* , Vol. 29(1), pp. 98–114.

Levinson, D. and Chen, W. (2005), Paving New Ground: A Markov Chain Model of the Change in Transportation Networks and Land Use, *in* D. Levinson and K. Krizek, eds, 'Access to Destinations', Elsevier Publishers.

Levinson, D. and Chen, W. (2007), "Area-Based Models of Highway Growth", *ASCE Journal of Urban Planning and Development* , Vol. 133(4), pp. 250–254.

Levinson, D., de Oca, N. M. and Xie, F. (2006), Beyond Business as Usual: Ensuring the Network We Want Is the Network We Get, Technical report, Mn/DOT 2006-36.

Levinson, D. and El-Geneidy, A. (2009), "The Minimum Circuity Frontier and the Journey to Work", *Regional Science and Urban Economics* , Vol. 39, Elsevier, pp. 732–738.

Levinson, D. and Karamalaputi, R. (2003*a*), "Induced Supply: A Model of Highway Network Expansion at the Microscopic Level", *Journal of Transport Economics and Policy* , Vol. 37, pp. 297–318.

Levinson, D. and Karamalaputi, R. (2003*b*), "Predicting the Construction of New Highway Links", *Journal of Transportation and Statistics* , Vol. 6, pp. 81–89.

Levinson, D., Xie, F. and Montes de Oca, N. (2007), 'Forecasting and Evaluating Network Growth', *Networks and Spatial Economics*, Accepted for publication.

Levinson, D. and Yerra, B. (2002), "Highway Costs and the Efficient Mix of State and Local Funds", *Transportation Research Record* , Vol. 1812, pp. 27–36.

Levinson, D. and Yerra, B. (2006), "Self Organization of Surface Transportation Networks", *Transportation Science* , Vol. 40, pp. 179–188.

Levinson, D. and Zhang, L. (2004), "Evaluating Effectiveness of Ramp Meters", *Assessing the Benefits and Costs of ITS* , Springer, pp. 145–166.

Levinson, D. and Zhang, L. (2006), "Ramp meters on trial: Evidence from the Twin Cities metering holiday", *Transportation Research Part A: Policy and Practice* , Vol. 40, Elsevier, pp. 810–828.

Levinson, M. (2006), *The Box*, Princeton University Press.

Lieberman, M. and Montgomery, D. (1988), "First-Mover Advantages", *Strategic Management Journal* , Vol. 9, JSTOR, pp. 41–58.

Lieberman, M. and Montgomery, D. (1998), "First-Mover (Dis) Advantages: Retrospective and Link with the Resource-Based View", *Strategic Management Journal* , Vol. 19, John Wiley & Sons, pp. 1111–1125.

Liebowitz, S. J. and Margolis, S. E. (1995), "Path dependence, Lock-in and History", *Journal of Law, Economics, and Organization* , Vol. 11, pp. 205–226.

Liljeros, F., Edling, C. R., Amaral, L. A. N., Stanley, H. E. and Aberg, Y. (2001), "The Web of Human Sexual Contacts", *Nature* , Vol. 411, pp. 907–908.

Lorz, O. and Willmann, G. (2005), "On the Endogenous Allocation of Decision Powers in Federal Structures", *Journal of Urban Economics* , Vol. 57, pp. 242–257.

Lowe, J. and Moryadas, S. (1975), *The Geography of Movement*, Houghton Mifflin Company.

Lowry, G. (1978), *Streetcar Man*, Lerner Publications Company Minneapolis.

Lowry, I. (1963), "Location Parameters in the Pittsburgh Model", *Papers and proceedings of the Regional Science Association* , Vol. 11, pp. 145–165.

Mackett, R. (1983), 'The Leeds Integrated Transport Model (LILT)', Supplementary Report 805.

Mackett, R. (1990), "The Systematic Application of the LILT Model to Dortmund, Leeds and Tokyo", *Transportation Reviews* , Vol. 10, pp. 323–338.

Mackett, R. (1991), "LILT and MEPLAN: A Comparative Analysis of Land-use and Transport Policies for Leeds", *Transportation Reviews* , Vol. 11, pp. 131–141.

Makadok, R. (1998), "Can First-Mover and Early-Mover Advantages Be Sustained in an Industry with Low Barriers to Entry/Imitation?", *Strategic Management Journal* , Vol. 19, John Wiley & Sons, pp. 683–696.

Marini, M. A. (2007), 'An Overview of Coalition & Network Formation Models for Economic Applications', Working Paper Series in *Economics, Mathematics, and Statistics, 2007/12*.

Marlette, J. (1959), *Electric Railroads of Indiana*, Indianapolis, Council for Local History.

Math Forum (2010), 'Famous Problems in the History of Mathematics', Retrieved on December 26, 2010.
  URL: *http://mathforum.org/isaac/problems/bridges1.html*

McDaniel, W. and Coley, M. (2004), "History of the Highway Trust Fund", *Transportation Research Record* , Vol. 1885, pp. 8–14.

Metropolitan Council (2007), 'Historical Transit Routes - Twin Cities Metropolitan Area'.
  URL: *http://www.datafinder.org/metadata/metrogis_regional_parcels.htm*

Metropolitan Council (2008), 'Regional Parcel Dataset - Seven County Twin Cities Metropolitan Area'.
  URL: *http://www.datafinder.org/metadata/metrogis_regional_parcels.htm*

Minnesota Department of Natural Resources (2003), Regionally Significant Ecological Areas (RSEA) 2003 Assessment Results, Technical report.
  URL: *http://www.dnr.state.mn.us/rsea/metro_assessment.html*

Minnesota Department of Transportation (2001), Metro Division Transportation System Plan, Technical report, MnDOT.

Minnesota Department of Transportation (2005a), Official Website.
  URL: *http://www.dot.state.mn.us/tda/html/faq.html*

Minnesota Department of Transportation (2005*b*), 2002 Freeway Volume-Crash Summary, Technical report, Office of Traffic, Security and Operations.
URL: *www.dot.state.mn.us/trafficeng/otepubl/CongestionReport-2004.pdf*

Mittal, S. and Swami, S. (2004), "What Factors Influence Pioneering Advantage of Companies?", *The Journal for Decision Makers* , Vol. 29(3), pp. 15–33.

Mohamed, A. (2007), Forecasting Transit Network Evolution, PhD thesis, University of Toronto.

Mohammed, A., Shalaby, A. and Mille, E. (2006*a*), "Empirical Analysis of Transit Network Evolution: Case Study of Mississauga, Ontario, Canada, Bus Network", *Transportation Research Record* , Vol. 1971, p. 5158.

Mohammed, A., Shalaby, A. and Mille, E. (2006*b*), 'Modeling the Supply of Public Transit: Using Artificial Intelligence to Model the Evolution of the Bus Transit Network', Working paper.

Mohring, H. and Harwitz, M. (1962), *Highway Benefits: An Analytical Framework*, Northwestern University Press.

Montes de Oca, N. (2006), Beyond Business as Usual: Ensuring the Network We Want is the Network We Get, Master's thesis, University of Minnesota at Twin Cities.

Montes de Oca, N. and Levinson, D. (2006), "Network Expansion Decision-making in the Twin Cities", *Transportation Research Record* , Vol. 1981, pp. 1–11.

Montgomery, M. R. and Bean, R. (1999), "Market Failure, Government Failure, and the Private Supply of Public Goods: The Case of Climate-controlled Walkway Networks", *Public Choice* , Vol. 99, p. 403437.

Mori, T. (2006), Monocentric versus Polycentric Models in Urban Economics, Discussion Paper 611, Kyoto University.
URL: *http://www.kier.kyoto-u.ac.jp/index.html*

Morrill (1965), "Migration and the Growth of Urban Settlement", *Lund Studies in Geography, Series B, Human Geography* , Vol. 26, pp. 65–82.

Mueller, D. (1997), "First-mover advantages and path dependence", *International Journal of Industrial Organization* , Vol. 15, Elsevier Science, pp. 827–850.

Murayama, Y. (1994), "The Impact of Railways on Accessibility in the Japanese Urban System ", *Journal of Transport Geography* , Vol. 2, pp. 87–100.

Nakicenovic, N. (1998), Dyanmics and Replacement of U.S. Transport Infrastructure, *in* 'Cities and Their Vital Systems–Infratructure, Past, Present and Future', National Academy Press, Washington DC.

Neuberger, H. (1971), "User Benefit in the Evaluation of Transport and Land Use Plans", *Journal of Transportation Economics and Policy* , pp. 52–77.

Newell, G. F. (1980), *Traffic Flow on Transportation Networks*, MIT Press, Cambridge.

Newman, M. (2001), "Clustering and preferential attachment in growing networks", *Physical Review E* , Vol. 64, APS, p. 25102.

Newman, M. (2003), "The Structure and Function of Complex Networks", *SIAM Review* , Vol. 45, pp. 167–256.

Oates, W. E. (1972), *Fiscal Federalism*, New York: Harcourt Brace Jovanovich.

Odlyzko, A. (2010), 'Collective Hallucinations and Inefficient Markets: The British Railway Mania of the 1840s'.
URL: *http://www.dtc.umn.edu/ odlyzko/doc/hallucinations.pdf*

Olson, R. (1976), *The Electric Railways of Minnesota*, Hopkins, MN, Minnesota Transportation Museum.

Oosterbeek, H., Sloof, R. and van de Kuilen, G. (2004), "Differences in Ultimatum Game Experiments: Evidence from a Meta-Analysis", *Experimental Economics* , Vol. 7, p. 171188.

Ortuzar, J. and Willumsen, L. G. (2001), *Modeling Transport*, John Wiley and Sons, LTD.

O'Sullivan, S. and Morrall, J. (2007), "Walking Distances to and from Light-Rail Transit Stations", *Transportation Research Record* , Vol. 1538 / 1996, pp. 19–26.

Parthasarathi, P., Levinson, D. and Karamalaputi, R. (2003), "Induced Demand: A Microscopic Perspective", *Urban Studies* , Vol. 40(7), pp. 1335–1353.

Payne-Maxie Consultants (1980), The Land Use and Urban Development Impacts of Beltways, Final report dot-os-90079, U.S. Dept. of Transportation and Dept. of Housing and Urban Development.

Peng, Z., Dueker, K. J., Strathman, J. and Hopper, J. (1997), "A Simultaneous Route-level Transit Patronage Model: Demand, Supply, and Inter-route Relationship", *Transportation* , Vol. 24, pp. 159–181.

Port of Hamburg (2005), Top 20 Ports: Container Throughput in TEU, Technical report, Port of Hamburg.

Powell, W. B. and Sheffi, Y. (1982), "Convergence of Equilibrium Algorithms With Predetermined Step Sizes", *Transportation Science* , Vol. 16(1), pp. 45–55.

Pred, A. (1966), *The Spatial Dynamics of U.S. Urban-Industrial Growth, 1900-1914*, Cambridge: The MIT Press.

Puffert, D. (2002), "Path Dependence in Spatial Networks: The Standardization of Railway Track Gauge", *Explorations in Economic History* , Vol. 39, Academic Press, pp. 282–314.

Rahman, Z. and Bhattacharyya, S. (2003), "First Mover Advantages in Emerging Economies: A Discussion", *Management Decision* , Vol. 41(2), pp. 141–147.

Redoano, M. and Scharf, K. A. (2004), "The Political Economy of Policy Centralization: Direct versus Representative Democracy", *Journal of Public Economics* , Vol. 88, pp. 799–817.

Riker, W. (1962), *The Theory of Political Coalitions*, Yale University Press.

Rimmer, P. (1967), "The Changing Status of New Zealand Seaports", *Annals of the Association of American Geographers* , Vol. 57, pp. 88–100.

Robertson, K. A. (1994), *Pedestrian Malls and Skywalks*, Ashgate Publishing, Limited.

Rodrigue, J., Comtois, C. and Slack, B. (2006), *The Geography of Transport Systems*, Routledge, London.

Rose, D. (1983), *The London Underground: A Diagrammatic History*, Douglas Rose.

Rosenstein, M. (2000), The Rise of Maritime Containerization in the Port of Oakland: 1950 to 1970, Master's thesis, Gallatin School of Individualized Study, New York University.

Safirova, E., Gillingham, K. and Houde, S. (2007), "Measuring Marginal Congestion Costs of Urban Transportation: Do Networks Matter?", *Transportation Research A* , Vol. 41 (8), pp. 734–749.

Samaniego, H. and Moses, M. E. (2008), "Cities as Organisms: Allometric Scaling of Urban Road Networks", *Journal of Transport and Land Use* , Vol. 1, pp. 28–39.

Schelling, T. C. (1978), *Micromotives and Macrobehavior*, New York, London: W.W.Norton.

Schmid, C. (1937), *Social Saga of Two Cities: An Ecological and Statistical Study of Social Treands in Minneapolis and St. Paul*, The Minneapolis Council of Social Agencies.

Schweitzer, F., Ebeling, F., Rose, H. and Weiss, O. (1998), "Optimization of Road Networks Using Evolutionary Strategies", *Evolutionary computation* , pp. 419–438.

Shannon, C. E. (1948), "A Mathematical Theory of Communication", *Bell System Technical Journal* , Vol. 27, pp. 379–423 and 623–656.

Shapiro, c. and Varian, H. R. (1998), *Information Rules: A Strategic Guide to the Network Econom*, Harvard Business School Press.

Sheffi, Y. (1985), *Urban Transportation Networks: Equilibrium Analysis with Mathematical Programming Methods*, Englewood Cliffs, NJ: Prentice-Hall, Inc.

Sole, R. V. and Valverde, S. (2004), "Information Theory of Complex Networks: On Evolution and Architectural Constraints", *Lecture Notes Physics* , Vol. 650, p. 189207.

Streeter, W. (1946), "Is the First Move an Advantage?", *Chess Review* , p. 16.

Suwansirikul, C., Friesz, T. L. and Tobin, R. (1987), "Equilibrium Decomposed Optimization: A Heuristic for the Continuous Equilibrium Network Design Problem", *Transportation Science.* , Vol. 21 (4), p. 261.

Taaffe, E., Gauthier, H. and OKelly, M. (1996), *Geography of Transportation.*

Taaffe, E., Morrill, R. L. and Gould, P. R. (1963), "Transportation Expansion in Underdeveloped Countries: a Comparative Analysis", *Geographical Review* , Vol. 53, pp. 503–529.

Taylor, B. (2004), The Geography of Urban Transportation Finance, *in* S. Hanson and G. Giuliano, eds, 'The Geography of Urban Transportation', New York: The Guilford Press.

Taylor, B. D. and Miller, D. (2003), Analyzing the Determinants of Transit Ridership Using a Two-Stage Least Squares Regression on a National Sample of Urbanized Areas, *Research Report Number 682*, University of California Transportation Center, Berkeley.

Tero, A., Takagi, S., Saigusa, T., Ito, K., Bebber, D. P., Fricker, M. D., Yumiki, K., Kobayashi, R., and Nakagaki, T. (2010), "Rules for Biologically Inspired Adaptive Network Design", *Science* , Vol. 327, pp. 439–442.

Tiebout, C. (1956), "A Pure Theory of Local Expenditures", *Journal of Political Economy* , Vol. 64(5), p. 416424.

Timmermans, H. (2003), The Saga of Integrated Land Use-Transport Modeling: How Many More Dreams Before We Wake Up?, *in* 'Proceeding of 10th International Conference on Travel Behaviour Research, Lucerne, August 10-15, 2003'.

Trusina, A., Maslov, S., Minnhagen, P. and Sneppen, K. (2004), "Hierarchy Measures in Complex Networks", *Physics Review Letter*, Vol. 92, pp. 178–702.

US Department of Commerce Bureau of the Census (1996), *Population of States and Counties of the United States: 1790 to 1990*, United States Department of Commerce, Bureau of the Census, Population Division.

Vaughan, R. (1987), *Urban Spatial Traffic Patterns*, London: Pion Ltd.

Verhoef, E. and Rouwendal, J. (2004), "Pricing, Capacity Choice, and Financing in Transportation Networks", *Journal of Regional Science*, Vol. 44(3), pp. 405–435.

Verhoef, E. T., Nijkamp, P. and Rietveld, P. (1996), "Second-best Congestion Pricing: The Case of an Untolled Alternative", *Journal of Urban Economics*, Vol. 40, pp. 279–302.

Vitins, B. J. and Axhausen, K. (2010), 'Patterns and Grammars for Transport Network Generation', 10th Swiss Transport Research Conference.

von Neumann, J. and Morgenstern, O. (1944), *Theory of Games and Economic Behavior*, Princeton University Press.

von Thünen, J. H. (1910), *Isolated State*, Jena: Gustav Fischer.

Warner, S. (2004), *Streetcar Suburbs: The Process of Growth in Boson, 1870-1900*, Harvard University Press.

Weidner, T. (1995), Hub Equilibrium in the US Airnet, Master's thesis, University of California at Berkeley.

Weingast, B. (1979), "A Rational Choice Perspective on Congressional Norms", *American Journal of Political Science*, Vol. 23, pp. 245–262.

Winston, C. (2000), "Government Failure in Urban Transportation", *Fiscal Studies*, Vol. 21(4), p. 403425.

Xie, F. (2005), The Evolution of Road Networks: A Simulation Study Based on Network Degeneration, Master's thesis, University of Minnesota at Twin Cities.

Xie, F. (2008), 'Validation of the Model of Network Degeneration: a case study of the Indiana Interurban Network', (08-0959) Presented at 87th Transportation Research Board (TRB) in Washington, DC, January 2008.

Xie, F. and Levinson, D. (2007*a*), "Measuring the Structure of Road Networks", *Geographical Analysis*, Vol. 39(3), pp. 336–356.

Xie, F. and Levinson, D. (2007*b*), "The Weakest Link: A Model of the Decline of Surface Transportation Networks", *Transportation Research part E*, Vol. 44E,1, pp. 100–113.

Xie, F. and Levinson, D. (2009), "Jurisdictional Control and Network Growth", *Networks and Spatial Economics*, Vol. 9(3), pp. 459–483.

Yamins, D., Rasmussen, S., D. and Fogel (2003), "Growing Urban Roads", *Networks and Spatial Economics*, Vol. 3, pp. 69–85.

Yang, H. and Bell, M. G. H. (1998), "Models and Algorithms for Road Network Design: A Review and Some New Developments", *Transport Reviews*, Vol. 18, pp. 257–278.

Yerra, B. and Levinson, D. (2005), "The Emergence of Hierarchy in Transportation Networks", *Annals of Regional Science* , Vol. 39, pp. 541–553.

Zhang, L. (2005), A Simulator of Network Growth for Network Economics and Policy Analysis, Master's thesis, University of Minnesota at Twin Cities.

Zhang, L. and Levinson, D. (2004), 'A Model of the Rise and Fall of Roads', Presented March 2004 at MIT Engineering Systems Symposium.

Zhang, L. and Levinson, D. (2005), "Road Pricing with Autonomous Links", *Transportation Research Record* , pp. 147–155.

Zhang, L. and Levinson, D. (2007), The Economics of Transportation Network Growth, *in* P. C. Millán and V. Inglada, eds, 'Essays on Transportation Economics', Springer, pp. 317–339.